Lancers and Dragoons

Lancers and Dragoons

Uniforms of the Imperial German Cavalry 1900-1914

by

R. J. Marrion

in collaboration with

D. S. V. Fosten and D. H. Hagger

ALMARK

ALMARK PUBLISHING CO. LTD.

First published 1975

ISBN O 85524 202 7 (hard cover edition)
ISBN O 85524 201 9 (limp cover edition)
Deposito Legal M-16981 1975

FRONT COVER: A Lieutenant of Dragoon Regiment No.23 in parade dress (left). An officer of Ulan Regiment No.17 in parade dress (right).

Printed in Spain by
Manufacturas Editoriales Litograficas S.A.
Poligono Industrial De Pinto (Madrid)
Carretera Fuenlabrada A Pinto K.M.21.
for the publishers Almark Publishing Co. Ltd.
49, Malden Way, New Malden, Surrey KT3 6EA.

ACKNOWLEDGEMENTS

The completion of this, the third and final part of the trilogy on the uniforms of German cavalry prior to the 1914-1918 war could not have been accomplished without the invaluable assistance of Messrs David Nash and Laurence Milner and the staff of the Imperial War Museum, London. Thanks are also due to Miss Frances Dimond of the Royal Archives, Windsor, and to Mr. R.G. Harris of Southsea, Hants, for allowing access to, and the use of, his unique and extensive photographic collection. Mr. Roy Belmont-Maitland of "Tradition", Piccadilly, has also to be thanked for allowing his fine collection of helmets to be photographed and Messrs Wallis and Wallis of Lewes, Sussex have also been generous in allowing me to use photographs from their catalogues. My German colleagues Herr Karl Redlin of Oberdresselndorf and Oberstleutnant Dr. Friedrich Herrman of Bonn are also to be thanked for their invaluable assistance in solving some of the more intangible problems which arose, and I must also offer my thanks to Mrs. Mary Nethercote for taking on the onerous and exacting task of typing the manuscript.

Finally, my special thanks are due to my two co-authors and illustrators Messrs Donald Fosten and Douglas Hagger without whose talents and unstinting help this work could never have been completed.

R.J. Marrion,
London, August 1974.

I: Dragoons

The Germans always considered the dragoon as a mounted infantryman. In early days they carried drums and infantry weapons and their name derives from the type of weapon which they originally carried. The dress of the German dragoon had always had some characteristics of the infantry, and the helmets and tunics followed the infantry pattern in many respects.

The light blue coat was first introduced for Prussian Dragoons in 1742, during the 2nd. Silesian War. By the time of the death of Frederick the Great (1786) there were twelve regiments in existence to which two additional units were added before the 1806 catastrophe of the Prussian Army at the Battle of Jena. In the subsequent re-organisation the Dragoons were reformed into six regiments. The 1st. Guard Dragoon Regiment was raised on the 21st February, 1815. Two further regiments were added later to bring the total to eight line regiments, plus the Guard regiment. By 1818, the list of dragoon regiments was as follows:

Regiment	Title	Facings	Buttons
	Garde Dragoner Regt.	Red	Yellow
1	Königen	Crimson	White
2	1 Westpreussisches	White	White
3	Litthauisches	Red	Yellow
4	2 Westpreussisches	Red	White
5	Brandenburgisches	Black	Yellow
6	Neumarkisches	Light Red	White
7	Rheinisches	Yellow	White
8	Magdeburgisches	White	Yellow

In 1819, four of these regiments became cuirassiers and the remainder were re-numbered from 1 to 4 with the titles (in numerical order) Litthauisches, Brandenburgisches, Neumarkisches, and Rheinisches. The facings were red, black, light red, and white respectively. In 1848, the colour of the coats became officially 'cornflower blue', the regimental facings being red, black, rose-red and white. The 4th. Regiment had two changes of facings between 1848 and 1858, in 1850 the facings for the regiment changed from white to 'Pompadour Red', and by 1858 the facings had again changed to 'Sulphur Yellow'.

In 1860, a second Guard Dragoon Regiment had been raised together with a further four line regiments, numbered 5 through to 8. The 2nd. Guard Dragoon Regiment wore red facings and had white buttons; the 5th. Regiment had red facings and white buttons; the 6th. Regiment black facings and white buttons; the 7th. Regiment rose-red facings and yellow buttons; and the 8th. yellow facings with yellow buttons. A further eight regiments were

Left: Prince George of Wales (later George V) wearing his uniform as Colonel-in-Chief of the 1st. Guard Dragoon Regiment in 1890 (reproduced by gracious permission of Her Majesty the Queen).

raised in 1866, numbered through from 9 to 16. The 9th. and 10th. Regiments had white facings, the 11th. and 12th. crimson, the 13th. poppy red, the 14th. black, the 15th. rose-red and the 16th. yellow. The 13th. to 16th. Regiments were granted the distinction of wearing a white piping on their collars and cuffs. Buttons were yellow for the 9th. 11th., 13th. and 14th. Regiments and white for the remainder.

Collars of the coats had been the same colour as the coat, i.e. cornflower blue, with only patches at the front of the collar in the regimental facing colour. However, in 1867, all regiments were ordered to have the collars of the jackets completely in facing colour, the 13th. to 16th. Regiments still retaining their white piping.

The Mecklenburg Regiments were brought into the numbered sequence of national regiments in 1867 taking the numbers 17 and 18, and, at the same time, the Oldenburg Dragoons came into the establishment as the 19th. Regiment. The Baden contingent were wearing light blue coats before entering into the national sequence of regiments and became the 20th., 21st. and 22nd. Regiments. They received red, light yellow, and black facings respectively, the latter piped with red. All had white buttons and were given white plumes to the helmet. When the Hessians came into the Convention army they also took into use the Prussian style uniform, but still adhered to their traditional dark green tunics. The regiments were numbered 23 and 24, the 23rd. Regiment having red facings, white buttons, and white Guard Lace on the collars and cuffs, and the 24th. white facings and white buttons. By 1871, the Württemburg cavalry had been re-organised as two Dragoon and two Ulan regiments. The Dragoons became the 25th. and 26th. Regiments with white and yellow facings respectively and with yellow and white buttons. The 25th. Regiment had Guard Lace on the collars and cuffs.

THE DRAGOON REGIMENTS

Title	Raised	Garrison	Army Corps
1. Garde Dragoner Regt. Königin Viktoria von Grossbritannien und Irland	21. 2.1815	Berlin	Guard
2. Garde Dragoner Regt. Kaiserin Alexandra von Russland	7. 5.1860	Berlin	Guard
Dragoner Regt. Prinz Albrecht von Preussen (Litthauisches) Nr.1	19. 4.1717	Tilsit	I
1. Brandenburgisches Dragoner Regt. Nr.2	24. 4.1689	Schwedt a.d. Oder	III
Grenadier Regt. zu Pferde Freiherr von Derfflinger (Neumarkisches) Nr.3	29.12.1704	Bromberg	II
Dragoner Regt. von Bredow (1. Schlesisches) Nr.4	7. 3.1815	Luben	V

Title	Raised	Garrison	Army Corps
Dragoner Regt. Freiherr von Manteuffel (Rheinisches) Nr.5	7. 5.1860	Hofgeismar	XI
Magdeburgisches Dragoner Regt. Nr.6	7. 5.1860	Mainz	XVIII
Westfälisches Dragoner Regt. Nr.7	7. 5.1860	Saarbrücken	XXI
Dragoner Regt. König Friedrich III (2. Schlesisches) Nr.8	7. 5.1860	Oels (Staff & 1st & 4th Squadrons) Kreuzberg(2nd Squadron) Bernstadt (3rd Squadron) Namslau (5th Squadron)	VI
Dragoner Regt. König Carl I von Rumanien (1. Hannoversches) Nr.9	25.11.1805	Metz	XVI
Dragoner Regt. König Albert von Sachsen (Ostpreussisches) Nr.10	27. 9.1866	Allenstein	XX
Dragoner Regt. von Wedel (Pommersches) Nr.11	27. 9.1866	Lyck	XX
Dragoner Regt. von Arnim (2. Brandenburgisches) Nr.12	27. 9.1866	Gnesen	II
Schleswig-Holstein Dragoner Regt. Nr.13	27. 9.1866	Metz	XVI
Kurmärkisches Dragoner Regt. Nr.14	27. 9.1866	Colmar	XV
Schlesisches Dragoner Regt. Nr.15	27. 9.1866	Hagenau	XV
2. Hannoversches Dragoner Regt. Nr.16	24. 3.1813	Luneburg	X
1. Grossherzoglich Mecklenburgisches Dragoner Regt. Nr.17	6.11.1819	Ludwigsburg	IX
2. Grossherzoglich Mecklenburgisches Dragoner Regt. Nr.18	8. 8.1867	Parchim	IX
Oldenburgisches Dragoner Regt. Nr.19	26. 4.1849	Oldenburg	X

Title	Raised	Garrison	Army Corps
1. Badisches Leib Dragoner Regt. Nr.20	23. 3.1803	Karlsruhe	XIV
2. Badisches Dragoner Regt. Nr.21	6. 1.1850	Bruchsal (Staff, 1st, 2nd, 3rd & 5th Squadrons) Schwetzingen (4th Squadron)	XIV
3. Badisches Dragoner Regt. Nr.22	6. 1.1850	Mülhausen	XIV
Garde Dragoner Regt. (1. Grossherzoglich Hessisches) Nr.23	6. 4.1790	Darmstadt	XVIII
Leib Dragoner Regt. (2. Grossherzoglich Hessisches) Nr.24	1.12.1859	Darmstadt	XVIII
Dragoner Regt. Königin Olga (1. Württembergisches) Nr.25	17.11.1813	Ludwigsburg	XIII
Dragoner Regt. König (2. Württembergisches) Nr.26	6.12.1805	Stuttgart-Canstatt	XIII

COLONELS-IN-CHIEF

Regiment	Colonel-in-Chief
1st. Guard	King George V of Great Britain
2nd. Guard	Czar Nicholas II of Russia
1st.	—
2nd.	—
3rd.	King Gustav V of Sweden
4th.	—
5th.	Prince Alfons of Bavaria
6th.	—
7th.	Prince Leopold of Bavaria
8th.	The Crown Princess of Germany
9th.	King Carol I of Rumania
10th.	—
11th.	—
12th.	Princess Eitel Friedrich of Prussia
13th.	—
14th.	Princess August Wilhelm of Prussia
15th.	Prince Ludwig of Bavaria
16th.	King Albert of Belgium
17th.	Grand Duke Friedrich Franz IV of Mecklenburg-Schwerin
18th.	The Grand Duchess of Mecklenburg-Schwerin
19th.	Grand Duke Friedrich August of Oldenburg
20th.	Grand Duke Friedrich II of Baden
21st.	The Grand Duchess of Baden
22nd	—
23rd.	Grand Duke Ernst Ludwig of Hesse
24th.	Czar Nicholas II of Russia
25th.	Emperor Wilhelm II of Germany and Prussia
26th.	King Wilhelm II of Württemberg

REGIMENTAL FACINGS AND BUTTON COLOURS

Regiment	Facings	Button Colour
Prussia		
1st. Guard*	Red	Yellow
2nd. Guard**	Red	White
1st.	Red	Yellow
2nd.	Black	Yellow
3rd.	Rose Red	White
4th.	Pale Yellow	White
5th.	Red	White
6th.	Black	White
7th.	Rose Red	Yellow
8th.	Yellow	Yellow
9th.	White	Yellow
10th.	White	White
11th.	Crimson	Yellow
12th.	Crimson	White
13th.	Red	Yellow
14th.	Black	Yellow
15th.	Rose Red	White
16th.	Yellow	White
Mecklenburg		
17th.*	Red	Yellow
18th.**	Black	White
Oldenburg		
19th.	Black	White
Baden		
20th.	Poppy Red	White
21st.	Light Yellow	White
22nd.	Black with red piping	White
Hesse		
23rd.**	Red	White
24th.	White	White
Württemberg		
25th.**	White	Yellow
26th.	Lemon Yellow	White

* Yellow "Garde-litzen" (gold for officers)
** White "Garde-litzen" (silver for officers)

The Grand Duchess of Mecklenburg-Schwerin in the uniform of Colonel-in-Chief of Dragoon Regiment No.18.

The Kaiser in the dragoon überrock. Note the sword hooked under the coat (R.G. Harris).

An Etatsmässige Wachtmeister wearing the mantel. Note the rank strips above and below the Guard lace on the collar patches.

RANKS AND DISTINCTIONS

General Officers: General-Oberst (no equivalent); General *(General der Kavallerie)*; Lieutenant-General *(General-Leutnant)*; Major-General *(General-Major)*.
Field Officers: Colonel *(Oberst)*; Lieutenant-Colonel *(Oberst-Leutnant)*; Major *(Major)*.
Captains and Subalterns: Captain *(Rittmeister)*; Lieutenant *(Ober-Leutnant)*; Second Lieutenant *(Leutnant)*.
Senior N.C.O.'s entitled to wear the officers' sword-knot:
Permanent Staff Regimental Sergeant-Major *(Etatsmässige Wachtmeister)*; Regimental Sergeant-Major* *(Wachtmeister)*; Squadron Sergeant-Major *(Vizewachtmeister)*; Acting Officer *(Offizier-Stellvertreter)*; Ensign/Officer Cadet *(Portepeefähnrich/Fahnenjunker)*.
N.C.O.'s without the officers' sword-knot:
Sergeant *(Sergeant)*; Corporal *(Unteroffizier)*†; Lance-Corporal *(Gefreite)*; One-Year Volunteer *(Einjahrig-Freiwillige)*; Re-engaged Volunteer *(Kapitulant)*.

* Trumpet-Major *(Stabstrompeter)* had same rank
† Trumpeter *(Trompeter)* had same rank

Dragoon officers wore two different patterns of shoulder ornament. For full-dress and ceremonial occasions they wore epaulettes, and, for other duties, shoulder cords. The epaulette was a shoulder board in the colour of the facings edged with silver lace striped in the state colours, and with a large metal half-moon or crescent in the button colour. The epaulette was retained on the shoulder by a small laced bridle, stitched on the shoulder of the coat, through which the epaulette passed, and by a threaded button at the collar end. The whole ornament was usually lined with cloth in the tunic colour. The rank was indicated by a system of stars and, for the senior ranks, also by fringing of bullion or metal wire fringe in the button colour. The shoulder cords had an underlay of the facing colour and a small frog which slipped under the shoulder bridle. At the other end, they were also retained by a button. Ranks were indicated by twisted silver and gold cords for General Officers, twisted silver for Field Officers, and straight silver cords for Captains and Subalterns. In all cases the silver cords were mixed with silk darts in the state colour.

Rank	**Epaulette Fringe**	**Stars**
Colonel-General*	Bullion	Three
General	Bullion	Two
Lieutenant-General	Bullion	One
Major-General	Bullion	Nil
Colonel	Thin metal	Two (one on either side of device, if applicable)
Lieutenant-Colonel	Thin metal	One (below device, if applicable)
Major	Thin metal	Nil
Captain	No Fringe	Two (one on either side of device, if applicable)
1st. Lieutenant	No Fringe	One (below device, if applicable)
2nd. Lieutenant	No Fringe	Nil

* General-Field-Marshal had crossed marshal's batons; a Colonel-General with the appointment of General-Field-Marshal had four stars (until 1911, crossed batons plus three stars)

Regimental Sergeant-Major: Gold or silver lace on the collar and cuffs, with a large heraldic button on each side of the collar. These flat collar buttons were gold or silver according to the button colour of the regiment and were engraved with the heraldic device of the state, e.g. the eagle in the case of the Prussian Regiments. In addition the Etatsmässige had a 1.6 cm. strip of similar gold or silver lace encircling each arm 7 mm. above the cuffs. On the greatcoat collar patches, two strips of worsted braid. These vertical strips of braid were in the state colours, i.e. white with a black central stripe for the

Prussians, white with light-blue for the Bavarians, white with green for the Saxons, white with red for the Hessians. The braid for Württemberg was red with a black stripe (for a regiment with red facings; black with a red stripe), and for Mecklenburg yellow with a half-red, half-blue centre stripe. In the case of regiments with guard lace, the strip was placed horizontally above the "litzen" for the Vize-wachtmeister and Unteroffizier and, additionally, below the "litzen" for the Wachtmeister. Officers' sword-knot.

Squadron Sergeant-Major: Gold or silver lace on the collar and cuffs plus a large heraldic button on each side of the collar. One strip of worsted braid on the greatcoat collar patches. Officers' sword-knot.

Acting Officers: Gold or silver lace on the collar and cuffs, with a large heraldic button on each side of the collar. Cockades, sabre and sword-knot of officers' pattern. Rank and file shoulder-straps but with an edging of the gold or silver collar/cuff lace and with the appropriate regimental devices of officers' pattern.

Ensign: Ranked senior to the longest serving Sergeant. After passing the prescribed examination the Ensign became a Degenfähnrich and then ranked senior to the Wacht-meister. Initially, the Ensign wore the distinctions of a Corporal, with the officers' sword-knot on the sabre. As a Degenfähnrich he wore the officers' sword and officers' cockades on the helmet and undress cap. From 1910, a special greatcoat was introduced for this rank. It had a single row of buttons down the front but was nevertheless cut double-breasted and closed by hooks and eyes. The collar patches were decorated as for the Squadron Sergeant-Major, the shoulder-straps were those of the rank and file, and the back of the garment was cut like an officer's paletot.

Sergeant: Gold or silver lace on the collar and cuffs, plus a large heraldic button on either side of the collar. Special N.C.O's sword-knot.

Corporal: Gold or silver lace on the collar and cuffs. Special N.C.O's sword-knot.

Lance-Corporal: A small heraldic button on each side of the collar. The Lance-Corporal was the lowest ranking non-commissioned officer and was normally in charge of the barrack room, but mounted guard with the troopers. The Lance-Corporal could be appointed by the Commanding Officer of the regiment after he had completed six months service.

The regimental band of the 3rd. Baden Dragoon Regiment No.22. This photograph was taken in the barrack square at Karlsruhe in about 1890.

Trumpeters: Normally ranked as Corporal but could rise to the rank of Sergeant or Squadron Sergeant-Major. In addition to their N.C.O.'s rank badges and the special trumpeter lace on collar and cuffs, they wore *schwalbennester* (bandsmen's wings) of the regimental facing colour, trimmed with gold or silver lace. The trumpet cords were of various colours according to the state from which the regiment originated. Trumpeter as a title is a misnomer; in fact, they were musicians capable of playing most of the wind instruments used by the military bands of the period.

Changes in titles of Ranks

In 1899, the following changes in title occurred:-

Offizieraspirant	became	Fahnenjunker
Portepeefähnrich	became	Fähnrich
Second Leutnant	became	Leutnant
Premier Leutnant	became	Oberleutnant

Two views of the other ranks' pattern pickelhaube with the parade plume attached. The helmet plate is the Dragoneradler (Imperial War Museum).

SPIKED HELMET *(Pickelhaube)*

All Dragoon Regiments wore polished black leather helmets with yellow or white metal (gilt or silvered for officers) fittings. The helmet was similar in pattern to the headdress worn by the German infantry of the period except that the front peak or visor was cut square instead of round. The exception to this rule were the 23rd. and 24th. (Hessian) Regiments who did have rounded front visors.

The crown and the front and rear visors were made of boiled black leather, highly polished and with a metal trim to the front visor. A spine or splint of metal ran from the back of the base plate of the spike to the edge of the rear visor. The front of the helmet was decorated with a plate which identified the regiment, consisting of either the Prussian Eagle or the State Arms. The rounded spike (fluted for the 17th. and 18th. Regiments) had an ovoid base plate, fixed to the crown with dome headed screws. For the officers the spike was taller and the base plate was cruciform and fitted with fluted star screws. For Parade Dress the spike was unscrewed and replaced by a tall fitment which took a pendant horsehair plume. The helmet had chin scales from circular bosses set high on the sides of the crown. Behind the

An other ranks' pattern pickelhaube from a Guard dragoon regiment with the parade plume attached. (Tradition).

A dragoon on manoeuvres. The pickelhaube is covered by a manöverkappe with a red band around it.

bosses were cockades. On the left side was the State cockade, and on the right the red white and black 'National' cockade. For the officers the cockades were deeply incised and the white sections were silvered. The 3rd. Regiment had the added distinction of small white metal (silvered for the officers) grenades over the chin scale bosses. On manoeuvres a reed-green duck cover was worn over the helmet. It was furnished with a red band.

Both Prussian Guard Regiments, the Baden Regiments, and the 25th. (Württemberg) Regiments had white hair plumes. All the remaining Regiments wore black plumes. Trumpeters of all Regiments wore red plumes.

Helmet plates and fittings were brass (gilt for the Officers) for the 1st. Guard; 1st.; 2nd.; 7th.; 8th.; 9th.; 11th.; 13th.; 14th.; 17th.; and 25th. Regiments. Fittings were in white metal (silvered for the Officers) for the 2nd. Guard; 3rd.; 4th.; 5th.; 6th.; 10th.; 12th.; 15th.; 16th.; 18th.; 19th.; 20th.; 21st.; 23rd.; 24th.; and 26th. Regiments.

The various types of helmet plates and the regiments which wore them are as follows:

Gardeadler: A crowned eagle looking to its right clutching the sword in its right talon and the sceptre in its left. The wings were outstretched with the top edges of the wings in a line with the eagle's beak. On its breast was the eight pointed Guard Star with the circlet lettered *SUUM CUIQUE* with sprays of laurel beneath the words and, in the centre of the circlet, the black eagle of Prussia. The tail feathers of this eagle were close and hung straight down between the legs.

Gardeadler without the star: The same eagle as described above but with the Star omitted and replaced with a five folded motto scroll over the wings and body, lettered *Mit Gott Für Koenig und Vaterland.* The tops of the wings of this eagle were slightly curved and the tail feathers were longer, reaching to the base of the talons and slightly curved to the right.

Dragoneradler: A crowned eagle looking to its right and with upswept wings (the tips of the longest, inside feathers curled inwards towards the head). The sword was held in the right talon and the sceptre in the left. The tail feathers were bunched but swept slightly out to fill the space between the legs. Across the wings and the breast was the three fold motto scroll lettered *Mit Gott Für Koenig und Vaterland.*

Badischer Greif: The griffon is crowned and has its head turned to look right. The lion's body is inclined to the left over a Baden shield and clutches in its right paw a spear-shaped sword with a cross hilt. The wings are upswept. Under the legs and shield a three fold scroll lettered *Mit Gott F. Fuerst U. Vaterland.*

Hessisches Löwen: A rampant crowned lion facing to its right clutching a cross hilted sword in its right paw and with two tails. Across the body were five red bands, the upper crossing the lower part of the muzzle and the lower the back paws. It was surrounded by a thick wreath of laurel on its right and oak on its left.

Württembergischer Wappen: The crest was a knight's helm facing front surmounted by a crown. The supporters were a crowned lion and a stag. The three fold motto scroll was lettered *Furchtlos und Treu.* The eight pointed Star over the shield had a centre bearing a Maltese style cross with a circlet around it lettered as the motto scroll.

The 1st. and 2nd. Prussian Guard Regiments had the Gardeadler in brass or white metal with the white 'Gardestern' on the breast. The 1st. Regiment had the Gardeadler without the Guard Star. The 2nd. Regiment had the Dragoneradler. From 1813 they had an oval white metal shield over the lower half of the body and tail feathers, the shield bearing the Prussian sceptre topped by a crown. For officers, the oval shield was blue enamel with a gilt sceptre and crown backed with red enamel. The 3rd. Regiment had the Gardeadler without the Star on its breast. The regiments numbering from 4 to 16 had the Dragoneradler. The 9th. Regiment had, on their plate, the honours "Peninsula, Waterloo and Gohrde" and the 16th. Regiment had the honour "Waterloo". The 17th. and 18th. Regiments had a Sunburst bearing the Mecklenburg Arms in the centre, the 17th had a yellow Sunburst with white metal Arms, and a white metal Sunburst with gilt Arms for the 18th. Regiment. The 19th Regiment had the Dragoneradler with a brass heraldic star on the breast, bearing the Arms of Oldenburg in white metal on it. The 20th., 21st. and 22nd. Regiments had the Baden heraldic Griffon in white metal clutching a shield in its talons. The 23rd. and 24th. Regiments had a white metal plate consisting of the crowned Hessian Lion within a wreath of laurel and oak leaves. The 25th. and 26th. Regiments had the full Arms of the Kingdom of Württemberg, in brass for the 25th. and in white metal for the 26th. Superimposed on the 26th. Regiment plate was the Star of the Order of the Crown of Württemberg.

Officers' helmet plate of the 1st. and 2nd. Prussian Guard Regiments — Gardeadler.

Helmet plate of Dragoon Regiments Nos. 23 and 24 — Hessisches Löwen.

Helmet plate of Dragoon Regiment No.26 — Württembergischer Wappen

Other ranks' helmet plate of Dragoon Regiment No.16 — Dragonadler.

Helmet plate of Dragoon Regiments Nos. 20, 21 and 22 — Badischer Greif.

Helmet plate of Dragoon Regiments Nos. 17 and 18 — Sunburst Star with the arms of Mecklenburg. (Imperial War Museum)

UNDRESS CAP *(Mütze)*

The dragoon undress cap was cornflower blue for all regiments except the Hessian Regiments, who wore a dark green cap. The cap band and piping were in regimental facing colour. On the cap band was a small round metal cockade in the state colours and directly above this, on the crown of the cap, a similar cockade in the national colours of black, white and red. The white area of the cockade was silver for officers and certain N.C.O.'s. In addition, Regiment No. 2 wore a small metal eagle (a very much smaller version of the helmet plate) in between the cockades.

Officers and senior N.C.O.'s had the cap fitted with a small black leather visor, while officers of Regiments 2, 6, 11, 12, 14, 18, 19 and 22 had the cap bands and pipings in fine velvet.

Officers of Contingent States serving on the Prussian Staff retained all regimental distinctions but changed the state cockade to the Prussian type. For example, a Mecklenburger or Bavarian wore his original uniform but with Prussian cockades on the full dress helmet or *Tschapska* and on the undress cap.

The officers' pattern undress cap.

The other ranks' pattern undress cap.

FULL DRESS TUNIC *(Waffenrock)*

The full dress tunic was a single breasted garment with eight buttons down the front, a standing and round fronted collar and point ended shoulder straps. It had 'Swedish' round cuffs with two buttons and the backs of the skirts were decorated with a slit and two vertical scalloped pocket flaps with two buttons on each. Instead of the upper third buttons two special button supports for the waist belt were provided.

The dragoon tunic was cornflower blue for all regiments except the 23rd. and 24th. (Hessian) Regiments, both of which retained dark green coats. The collar, cuffs and the point-ended shoulder straps of the coat were in the regimental facing colour and there was a piping of the same colour down the leading edge of the coat and along the edges of the pocket flaps on the skirts. The collars were piped cornflower blue.

Both of the Prussian Guard Regiments had Guard Lace bars *(Garde-litzen)* on the collars and cuffs (yellow for the 1st. Regiment and white for the 2nd.) and the collars of their tunics were slightly higher, cut square and hooked close in front. The 23rd. (Hessian) Regiment also had these tight square fronted collars. The 13th., 14th., 15th. and 16th. Regiments had the

Prince Albert of Schleswig-Holstein in the uniform of Dragoon Regiment No.23. (reproduced by gracious permission of Her Majesty the Queen).

The other ranks' full dress tunic of Dragoon Regiment No.3. Note the early pattern shoulder strap and the fechterabzeichen (David Nash).

traditional white piping on the collars and cuffs and the 22nd. (Baden) Regiment had red pipings on collars, cuffs, pockets, and down the front. The 17th., 18th., 23rd. and 25th. Regiments also had Garde-litzen on the collars and cuffs in the button colour. The shoulder straps were decorated with the regimental number or the regimental device. The numbers or devices were yellow on all shoulder straps except for those regiments with yellow or white facings in which case the numbers or devices were red. The 23rd. Regiment had red facings and white devices.

The colour of the officers' tunics was a much lighter shade and could be said to be 'pale sky blue'. The facings of the 2nd., 6th., 11th., 12th., 14th. and 18th. Regiments were velvet. The 11th. and 12th. Regiments also had velvet pipings. The facings of these latter regiments were officially crimson but for the officers of the regiments the colour was a very dark rich tone indeed, almost violet.

Regiment	Distinction
1st Guard *	Cypher VIR surmounted by a crown; in yellow for other ranks, gold for officers.
2nd Guard *	Cypher A surmounted by a crown; in white for other ranks, silver for officers.
1st.	Numeral 1; in yellow for other ranks, gold for officers.
2nd.	Numeral 2; in yellow for other ranks, gold for officers.
3rd.	Cypher GV surmounted by a crown; in yellow for other ranks, gold for officers.
4th.	Numeral 4; in red for other ranks, gold for officers.
5th.	Numeral 5; in yellow for other ranks, gold for officers.
6th.	Numeral 6; in yellow for other ranks, gold for officers.
7th.	Numeral 7; in yellow for other ranks, gold for officers.
8th.	Crowned cypher FR with a Roman III below; in red for other ranks, gold for officers.
9th.	Crowned cypher C; in red for other ranks, gold for officers.
10th.	Crowned cypher AR; in red for other ranks, gold for officers.
11th.	Numeral 11; in yellow for other ranks, gold for officers.
12th.	Numeral 12; in yellow for other ranks, gold for officers.
13th.	Numeral 13; in yellow for other ranks, gold for officers.
14th.	Numeral 14; in yellow for other ranks, gold for officers.
15th.	Numeral 15; in yellow for other ranks, gold for officers.
16th.	Numeral 16; in red for other ranks, gold for officers.
17th. *	Crowned cypher FIII; in yellow for other ranks, gold for officers.
18th. *	Crowned cypher A; in red for other ranks, gold for officers.
19th.	Crowned cypher A; in red for other ranks, gold for officers.
20th.	A crown; in yellow for other ranks, gold for officers.
21st.	Numeral 21; in red for other ranks, gold for officers.
22nd.	Numeral 22; in yellow for other ranks, gold for officers.
23rd. *	Crowned cypher L; in white for other ranks, gold for officers.
24th.	Crowned cypher NII; in red for other ranks, gold for officers.
25th. *	Crowned Gothic cypher O; in red for other ranks, gold for officers.
26th.	Crowned Gothic cypher W; in red for other ranks, gold for officers.

* These regiments had *Garde-litzen* on the collars and cuffs in the button colour.

Waffenrock of a dragoon Unteroffizier.

Waffenrock of a dragoon Etatsmässige Wachtmeister.

Shoulder strap ornaments:
N.C.O.'s and troopers' shoulder strap ornaments. *Top row:* 1st. Guard Regiment; 2nd. Guard Regiment; Grenadiers zu pferde. *2nd. row:* Regiment No.8; Regiment No.9; Regiment No.10. *3rd. row:* Regiment No.17; Regiment No.18; Regiment No.19. *4th. row:* Regiment No.23; Regiment No.24; Regiment No.25. *5th. row:* Regiment No.26.

A dragoon officer wearing the überrock with the pouch belt and feldhinde over the coat.

The officers' pattern litewka.

OFFICERS' FROCK COAT *(Überrock)*

The frock coat was a double breasted garment in the colour of the officers' full dress tunic. It had two rows of six buttons down the front and a high collar with a rounded front. The cuffs were very deep and of a plain round design. The rear of the coat was decorated with a slit to waist level and two vertical pockets each fitted with two buttons. Unlike the three point scalloped flaps of the full dress coat these flaps were plain, broadening from the top towards the central slit in the centre each forming a long triangular shape with the base at the bottom. The collar, the lining to the breasts, the piping around the cuff edges and the edges of the pocket flaps were in the facing colour. The coat reached to the knee and was provided with hooks and eyes so that the skirts could be turned back when the officer was mounted. The sword was worn under the coat and suspended on a short strap and clasp.
Officers of the 23rd. and 24th. (Hessian) Regiments had dark green frock coats with collars, breast linings and pipings in the facing colour.

WORKING TUNIC *(Litewka)*

This grey coat was introduced for dragoons in 1897 to replace the former dark blue pattern. The word 'Litewka' derives from the Lithuanian language and was a garment which formed part of the country's national costume. The N.C.O.'s had been issued with the garment in 1895 but the officers did not discard their blue working tunic until 1903.

Officers Pattern: A double breasted light grey tunic made from doeskin (a fine cloth resembling undressed deer leather) with two rows of six buttons down the front, deep round cuffs, a stand and fall collar and horizontal pockets set in the sides of the skirts with external flaps. The front and lower edge of the collar, the leading edge of the front and the edge of the cuffs were piped. On each collar front was an oblong patch of cornflower blue cloth piped with facing colour. The coat pipings were also in facing colour with the exception of the 13th., 14th, 15th. and 16th. Regiments which had their traditional white pipings. The rear of the coat was quite plain.

The 23rd. (Hessian) Dragoons had a dark green collar patch piped with red and red piping on the coat front and cuffs. The 24th. (Hessian) Dragoons had dark green collar patches edged white and white piping on the coat. The 22nd. (Baden) Dragoon Regiment also had the collar patches piped with red. Regiments with black or crimson facings had velvet collar patches.

Photographs indicate that the officers' litewkas varied greatly in colour, some appearing almost white.

For General Officers the pattern was generally the same as worn by officers, but the pipings were poppy red, and the jacket had poppy red breast linings.

N.C.O.'s and Dragoons

A single breasted grey coat made from heavy melton cloth with a fly-front. The collar was the folded down pattern and had facing colour patches. The 13th., 14th., 15th. and 16th. Regiments had the rear edge of the patch only trimmed with white. Shoulder straps were cornflower blue with numbers or cyphers in the colours as worn on the full dress tunic. Before 1900 the numerals and cyphers were all yellow.

N.C.O.'s wore their braid strips in the state colours vertically on these patches and had gold or silver lace on the collar fronts and along the bottom edges. Where applicable, they also wore the heraldic button on the sides of the collar. The Sergeant-Major had, in addition, a treble chevron in gold or silver lace on the upper left arm. The coat was slit to the waist at the rear and had two buttons at waist level.

OFFICERS' SUMMER TUNIC *(Sommerock)*

For the summer half year officers could wear a white single breasted undress tunic, closed by eight regimental buttons down the front. It had no coloured facings or pipings, the cuffs were plain, and the shoulder cords were sewn directly to the coat without bridles.

OFFICERS' CLOAK *(Umhang)*

In 1889, a grey cloak was authorised. Reaching to well below the knee, the cloak could be worn alone or over the paletot. It had a fly-front and two inside pockets. The collar of the cloak was coloured differently when turned up or folded down, in the same manner as the paletot. It could be fitted with a detachable hood for foul weather, the hood being buttoned on either side under the folded down collar, or inside it. Photographs show the cloak to have been a voluminous garment which well covered the officer's knees and lay well over the horse's quarters behind the saddle.

OFFICERS' GREATCOATS *(Mantel/Paletot)*

Officers of dragoon regiments wore two patterns of light grey overcoat. Both garments were double breasted and had two rows of six buttons down the fronts and a 20″—28″ slit at the rear which could be closed by three or four small horn buttons. The two coats differed in that the mantel was primarily the overcoat worn for mounted duties and the paletot was worn on foot.

When not on duty the officer had the choice of either garment but, unless the weather was very bad, they normally wore the paletot. The bandolier and pouch were worn over the coat and the swordbelt under it. The sword was suspended on a short strap and clasp.

The collar of both garments was designed to be worn folded down or up and the colours differed on each side. The collars were cornflower blue when folded down, dark green for the Hessian Regiments. When turned up the collars were in the facing colour. The 13th., 14th., 15th. and 16th. Regiments had their traditional white pipings visible when the collar was worn up or down.

Mantel: A very long coat reaching to within two inches of the ankle. The coat fitted close at the neck and the deep collar, when turned up covered the ears. The coat had a deep slit at the back and had the same half belt and side and vertical pockets as the paletot. The collar had two lion heads and a short chain on the underside of the collar to facilitate closing it about the neck when turned up.

Paletot: A shorter, very elegant pale grey overcoat with a low but deep collar which when turned up reached to the lobe of the ear. The collar was cut so that the collar of the tunic worn beneath it showed above it when folded down. The coat reached to approximately nine and a half inches above the instep but photographs confirm that there was a great variety in the length of this garment. The back of the coat had a central slit and a horizontal half belt with one button and a vertical scalloped pocket flap on either side each with three buttons. The collar was fitted with two hooks and eyes to close the collar when turned up and there were two tabs at collar level stitched into the insides of the breasts one with a button the other with a buttonhole. There were in addition two pockets set in the sides of the skirts level with the lower buttons.

General Officers wore the paletot with special poppy red linings to the breasts and similar red pipings on the cuffs, pockets and half belts.

The General Officers' pattern paletot with the poppy red lining.

Dragoon N.C.O. (left) and infantry wearing the mantel.

OTHER RANKS' GREATCOATS *(Mantel)*

The other ranks wore an ankle length, single breasted thick grey melton coat with a slit at the back to facilitate its use for mounted duties. The coat had six buttons in the regimental colour down the front and a deep, folded down collar with rectangular cloth patches in the facing colour. These collar patches had a white strip along the rear edge for the 13th., 14th., 15th. and 16th. Regiments. Guard Lace was worn on these patches as on the full dress tunic. Shoulder straps on the overcoat were also as worn on the full dress tunic. N.C.O.'s wore the braid strips in the state colours on these collar patches; in the case of the regiments with Guard Lace, the braid strips were worn parallel with the bars and above them. In the case of Sergeant-Majors with two braid strips, one appeared above and the second was placed beneath the lower Guard Lace bar. Corporals had an overcoat button on each side of the underside of the collar, and the more senior N.C.O.'s, the heraldic state button. The 22nd. Regiment had black collar patches edged all round with red piping (as worn on the shoulder strap).

FATIGUE AND STABLE DRESS *(Drillichsachen,*

For fatigue duty in the barracks, drills, and for mucking-out stables, etc., the dragoon troopers and N.C.O.'s wore a loose greyish-yellow drill coat with a low standing collar and six small horn buttons down the front. N.C.O.'s wore a narrow striped braid band along the top and front edge of the collar, troopers a small cord of the state colours along the bottom of the collar, and the one-year volunteers a cord on the left shoulder, also in the state colours. The Etatsmässige Wachtmeister had an additional strip of braid around the sleeve at approximately cuff depth.

RIDING BREECHES *(Reithosen)*

Blue-black mixture breeches (almost black) were worn by the other ranks of all the dragoon regiments. Officers' breeches were universally dark-blue.

RIDING BOOTS *(Reitsteifel)*

Black leather boots reaching to just below the knee cap in front and cut to approximately 6cm. lower at the rear. There were no hobnails in the soles. N.C.O.'s and troopers' spurs were straight and the officers' curved. With the long trousers officers wore lighter and shorter style boots with spurs.

TROUSERS *(Tuchhosen)*

For walking-out and undress, very dark blue-black trousers were worn. These trousers were worn during the winter months of the year and were supported on wide canvas braces. They had poppy red stripes down the outside seam of each leg, except in the case of the 3rd., 7th. and 15th. Regiments, which had rose red piping, whilst the 11th. and 12th. Regiments had crimson piping and the 24th. Regiment, white. Photographs suggest that the officers' trousers were of narrower cut than those of the other ranks. Officers also wore the dark blue-black trousers for some mounted duties. Light boots with screwed on spurs were worn beneath the trousers.

During the summer months, between May and October, the N.C.O.'s and troopers were authorised to wear long trousers in white linen. Officers wore similar trousers for summer levees, during the ball season, in undress, and off duty. They were not worn on horseback.

GALA TROUSERS *(Galahosen)*

For State gala occasions and balls, etc., gala trousers were worn by officers. These were of cornflower blue for all regiments except Nos. 23 and 24. They carried double broad stripes, with a narrow piping between, in the regimental facing colour down the outside of each leg. They were velvet for Regiments 2, 6, 11, 12, 14, 18, 19 and 22. The last-named regiment differed in having the central piping in red whilst the outer edge of the black stripes was also piped in red. The Hessian regiments, Nos. 23 and 24, had black gala trousers, the 23rd. with red stripes and the 24th, white.

A group of dragoon officers wearing the mantel or the paletot with dress sashes. Note that the Adjutant on the right of the picture is not wearing a waist-belt or pouch-belt.

OFFICERS' DRESS SASH *(Schärpe)*

The officers' sash was in silver or gold thread worked with 5 cm. wide silk stripes in the state colours, and with silver or gold knots and tassels.

The sashes for the various contingents were as follows:

Prussia:	Silver, with two black stripes.
Bavaria:	Silver, with two light blue stripes.
Saxony:	Silver, with two green stripes (three for General Officers).
Württemberg:	Silver, with two black stripes, each of which had a poppy red stripe immediately below and adjoining.
Baden and Oldenburg:	Silver, with two 5 cm. wide black stripes and a 3 cm. wide red stripe in between.
Mecklenburg:	Gold, with cornflower blue, yellow and poppy red stripes.
Hesse:	Silver, with three red stripes.

Officers' undress waist-belts:

Prussia

Baden

Bavaria

Württemberg

Saxony (General officer range with three stripes)

Hesse

Mecklenburg-Schwerin

Mecklenburg-Strelitz

OFFICERS' UNDRESS WAIST-BELT *(Feldbinde)*

These belts were introduced in 1896, and were worn for orders of dress when the sash was not worn. The belts were approximately 5 cm. wide and were made in silver or gold woven thread with silk stripes worked in. They were closed by a locket and clasp at the front and had a slide in similar colours to the belt, one on either side of the clasp. The colours for the various contingents were as follows:

Prussia: A silver belt with two black stripes, the locket and clasp silver or gold according to button colour of the regiment. The locket design was the cypher WRII with a crown over, the clasp encircled in a closed wreath of laurel leaves.

Bavaria: A silver belt with two pale blue stripes, the locket and clasp silver or gold according to button colour. The locket design was the Bavarian crown, the clasp encircled in a closed wreath of laurel leaves.

Saxony: A silver belt with two (three for General Officers) pale green stripes, the locket and clasp silver or gold according to button colour. The locket design was a Gothic A with a crown over, the clasp encircled in a closed wreath of laurel leaves.
Württemberg: A silver belt with two black/red stripes, the locket and clasp silver or gold according to button colour. The locket design a Gothic W with a crown over, the clasp encircled in a closed wreath of laurel leaves.
Baden: A silver belt with one thin red stripe between two black stripes, the locket and clasp silver or gold according to regimental button colour. The locket design was the WRII and crown cypher as for the Prussians, the clasp completely encircled with a closed wreath of laurel leaves.
Oldenburg: The same belt as the Baden regiments.
Mecklenburg: A gold belt with, from top to bottom, a cornflower blue, yellow, and dark red stripe, the locket and clasp silver or gold according to button colour. The locket design was a Gothic FF cypher with a crown over, the clasp completely encircled with a closed wreath of laurel leaves.
Hesse: A silver belt with three red stripes, the locket and clasp silver or gold according to regimental button colour. The locket design was an EL Gothic cypher with a crown over, the clasp completely encircled with a closed wreath of laurel leaves.
Musikmeisters: Wore a special red Moroccan leather belt faced with facing colour cloth and trimmed along both edges with N.C.O. pattern lace of the button colour. The locket design was a Lyre with swords and the clasp was encircled by a wreath of laurel leaves. The locket and clasp was silver or gold according to button colour.

A group of dragoon officers wearing undress waist-belts.

POUCH-BELT *(Kartusche-Bandelier)*

For N.C.O.'s and troopers the pouch-belt was whitened leather, unless otherwise stated. Prussian officers had, universally, a belt of the tunic colour overlaid with lace in the button colour, the tunic colour showing as an edging on either side.

Officers of the 17th. Regiment had a belt of the tunic colour overlaid with lace in the button colour, as for the Prussian regiments, but the front was decorated with chains and signal pipes. The double chains were looped down from a cast lion's head to an oval silver plate which supported the tiny signal pipes or whistles. The 18th. Regiment had a similar pouch-belt, but covered in silver lace.

The officers of the 23rd. and 24th. Regiments wore red Moroccan leather pouch-belts for full-dress and parade. The belts were overlaid with silver lace, with a red edging showing on either side. The silver lace was divided along the length of the belt into three parts by red stripes, and was further embellished with lion's heads and signal whistles as worn by the 17th.

The 1st. Prussian Guard Dragoon Regiment on parade, showing the position of the pouch and the full dress parade shabraque.

The Grand Duke of Hesse wearing überrock, feldbinde and undress pouch-belt.

and 18th. Regiments. This belt was worn for all duties when the dress pouch was obligatory. For all other duties a black Moroccan leather belt was worn. This undress belt had a row of studs or silver domed buttons down the centre. The belts had special silver buckles, slides and tips at the back, and were fastened to the pouches by hooks known colloquially as "Attrappe".

The officers of the 25th. Regiment were distinguished by bandoliers which were in the tunic colour, overlaid with silver lace, and decorated with double silver chains looped from an upper oval plate. This plate was silver plated and had a gilded shield decorated with the WII cypher, with a crown resting on its top edge. The lower oval plate was also silver plated and had a similar shield but embellished with the cypher of the kings of Württemberg. The double silver chains were fastened to the lower side of the upper plate by a ring and looped down to a small tube on the lower side of the lower plate, into which the signal pipe or whistle was thrust. N.C.O.'s and troopers of this regiment had white pouch-belts but decorated with similar ornaments; the upper plate of German silver with brass WII and crown, the lower in similar coloured metal. The pipe and chains were German silver also.

Dragoon officers' pouch belts:

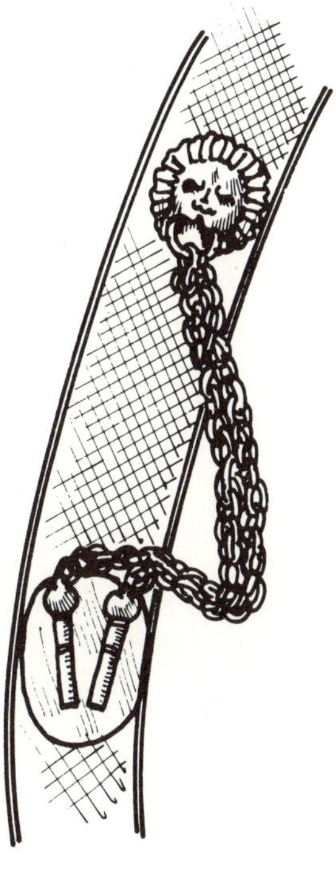

Regiment No.17

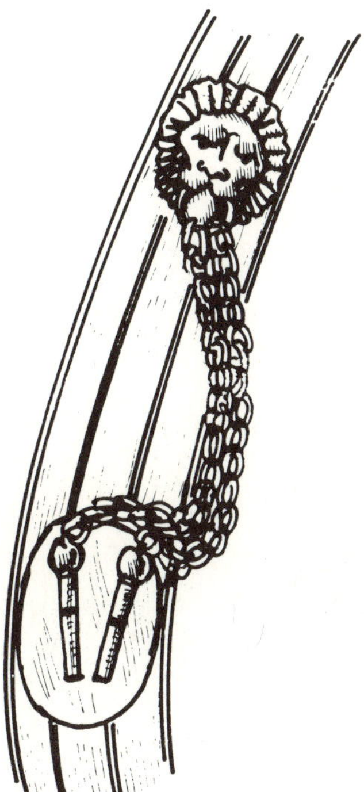

Regiments Nos. 23 and 24

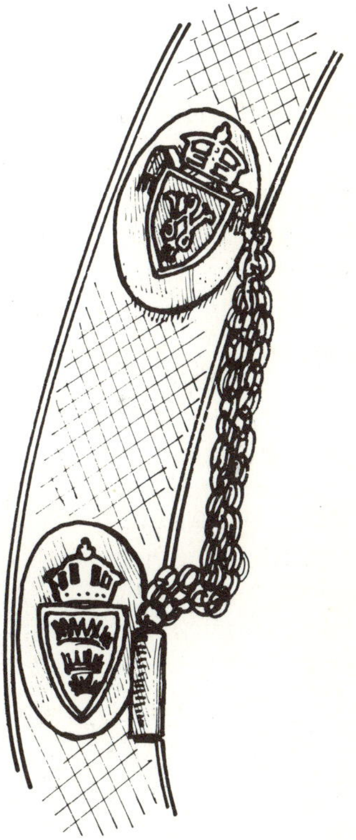

Regiment No.25

POUCH *(Kartuschkasten)*

The pouch-belt decorations of the German dragoon regiments were very varied and it is, perhaps, simpler to describe them in the form of a chart, as shown below:

	Pouch Ornament			**Pouch Ornament**	
Regt.	**Other Ranks**	**Officers**	**Regt.**	**Other Ranks**	**Officers**
1	A	A	14	A	A
2	A	B	15	A	A
3	C	C	16	A	A
4	A	A	17	F	G
5	A	A	18	F	G
6	A	A	19	A	H
7	A	A	20	M	I
8	A	D	21	M	I
9	A	E	22	M	I
10	A	A	23	J	J
11	A	B	24	J	J
12	A	A	25	—	K
13	A	A	26	L	L

The Prussian Guard regiments had pouch ornaments consisting of the Star of the Order of the Black Eagle. The 1st. Regiment had brass badges (gilt for officers), and the 2nd. Regiment had white metal (silver for the officers).

Legend

A: A brass (gilt for officers) circular plate with the Prussian eagle impressed in the centre upon a trophy of arms, drums and flags. The rim of the plate was raised. **B:** A gilded WR cypher with a crown above. **C:** A brass or gilt Star with a brass or gilt flaming grenade set on at angles in each corner. **D:** A gilded FR cypher. **E:** A gilded C cypher with a crown above. **F:** A Mecklenburg Star in brass. **G:** A silver plated pouch, the lid edged with gilt and with a gilt Mecklenburg Star in the centre. **H:** A gilded Gothic WR cypher with a crown above. **I:** A gilded FWR cypher with a crown above. **J:** N.C.O.'s and troopers wore a pouch ornament consisting of a brass L with a crown over. The officers wore a similar design but in uncoloured tooled leather. For parades, orderly duties with the Grand Duke, courts martial, church parades and guard duties, officers wore a special pouch which consisted of the same cypher and crown but in silver, surrounded by a gilded half open oak and laurel wreath and with the edge of the pouch decorated with a gilt trim. **K:** Officers of this regiment had a gilded Gothic KR and crown ornament. Note that the N.C.O.'s and troopers had plain pouches. **L:** The Star of the Order of the Crown of Württemberg in silver, with an enamelled centre for the officers, and in white metal for the other ranks. **M:** Other ranks of these three regiments had circular white metal plates impressed with the Baden Griffon clutching a shield in its front paws all on a trophy of arms.

Pouch ornaments:

All ranks of Prussian Guard Dragoon Regiments.

All ranks of Dragoon Regiments Nos. 1, 4, 5, 6, 7, 10, 12, 13, 14, 15 and 16, also other ranks of Dragoon Regiments Nos. 2, 8, 9, 11 and 19.

Officers of Dragoon Regiments Nos. 2 and 11 and Ulan Regiments Nos. 13, 19 and 20.

All ranks of Dragoon Regiment No.3.

Other ranks of Dragoon Regiments Nos. 17 and 18.

Officers of Dragoon Regiment No.19

Other ranks of Dragoon Regiments Nos. 20, 21 and 22.

All ranks of Dragoon Regiment No.26.

Fret-cut arms in the basket guard of the M89 sabre:

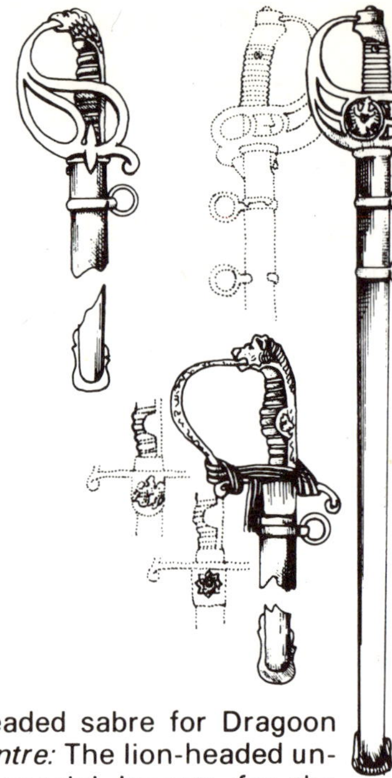

Left: The eagle-headed sabre for Dragoon Regiment No.2. *Centre:* The lion-headed undress sabre with special langets for the Guard Regiments and Dragoon Regiment No.1. *Right:* The M89 sabre.

SWORDS *(Säbel)*

Prussian Officers were armed with the M52/79 sabre, with a gilded basket guard. For undress duties, they carried a light sword *(stichdegen)* with a single stirrup guard and with a lion's head pommel. The Guard Regiments had small Guard Stars on the langets. N.C.O.'s and troopers were armed with the M89 pattern sabre with a brass guard bearing the state arms on the basket. Officers and Sergeant-Majors of the 1st. Regiment were also armed with the M52 sabre but with a silver-plated eagle on the side of the basket guard. For undress the officers wore the lion's head sword but with the silver eagle on the langets.

Officers and Sergeant-Majors of the 2nd. Regiment were armed with French Pattern cavalry swords which had a three branch bar guard with an eagle's head pommel. The grip was sharkskin, bound with silver wire. The 17th. and 18th. Regiments carried Prussian sabres, the officers using the lion's head sword for undress duties.

Officers of the Baden 20th., 21st. and 22nd. Regiments carried Austrian pattern swords with fretted silver guards, and the Hessian officers carried similar weapons. For undress duties, the Prussian lion's head stichdegen was carried.

The Württemberg 25th. and 26th. Regiments carried an Austrian pattern sword also, but with the state shield worked into the fretted basket guard.

OFFICERS' SWORD-KNOT *(Portepee)*

These were made of metallic braid and worked in coloured silks in the state colours.
Prussia: A black leather strap worked with three silver stripes; a black plaited slide worked with silver cord, and a silver and black worked stalk terminating in a silver acorn with a black silk core.
Bavaria: A crimson leather strap, faced with silver lace worked with two light-blue stripes. Over the strap, a silver slide, the strap joined to the stalk by a flattened split ring of silver. The stalk and tassel in silver cord worked through with light-blue silk. The tassel silver with a light-blue velvet core. A simple silver cord undress knot was also used.
Saxony: A black leather strap, edged with silver cord and having three stripes; the centre one green silk, the two outer stripes silver. The stalk and acorn silver cord with green silk worked in, the acorn having a green silk core.
Baden and Oldenburg: The Prussian pattern sword-knot, but with a narrow red silk stripe in the centre of the strap and a red silk core to the acorn.
Hesse: A black leather strap with a silver cord edge and a red silk stripe between. Stalk and tassel silver cord, worked through with red silk; the tassel with a red silk core.
Württemberg: A black leather strap with silver cord on each edge and a red silk stripe between. Stalk and tassel silver, the latter with a red and black silk core.
Mecklenburg: The Prussian pattern sword-knot but in gold cord with a red, yellow, and light-blue silk core to the acorn.

SWORD-KNOTS FOR N.C.O.'s AND TROOPERS *(Troddeln/Portepee)*

The sword-knot contributed a significant detail to the German uniform and was designed to give a great deal of information as to the rank and squadron identity of the wearer. The sword-knot comprised a strap, a slide, a 'crown', and a tassel.

Squadron	1st.	2nd.	3rd.	4th.	5th.
Strap	Red leather	Red leather	Red leather	Red leather	Red leather
Slide	Plaited red leather	Plaited red leather	Plaited red leather	Plaited red leather	Plaited red leather
Crown	White	Red	Yellow	Light blue	Green
Tassel	White	White	White	White	White

Sergeants and Corporals: The strap and slide was the same as the troopers', but the crown and tassel differed. The crown was made of white wool, mixed with the state colours; black for Prussia, light-blue for Bavaria, green for Saxony, red and black for Oldenburg, Baden and Württemberg, and light-blue, yellow and red for Mecklenburg. The fringe also differed in that it was mixed with the state colours. The Bavarians had a light-blue core to the white fringe, and the Hessians a red and white cord with a red core. The Mecklenburg tassels were entirely red, yellow, and light-blue. Württemberg entirely red mixed with black.

State patterns of Kapitulanten sword knots:

Prussia | Bavaria | Saxony | Hesse | Mecklenburg | Brunswick | Württemberg

Prussia | Bavaria | Saxony | Hesse | Mecklenburg | Brunswick | Württemberg

Officers' sword knots:

Left: The pattern used by officers of all states except Bavaria. *Right:* The Bavarian pattern.

Regimental Sergeant-Majors, Squadron Sergeant-Majors and Ensigns: All wore the officers' pattern metallic cord and silk sword-knot (see separately for full description).

Re-enlisted men: These personnel were distinguished by a special honour knot, which consisted of the leather strap and slide of the troopers but with a crown and tassel of a special shape and design, manufactured in cord in the state colours.

LANCE *(Lanze)*

All Corporals and other ranks carried the hollow steel shafted lance with a cruciform spike, as in the other arms of the cavalry. The lance pennons were coloured as follows:-
Prussia: Black over white pennon; Corporals carried an entirely white pennon with a black heraldic Prussian eagle thereon.
Mecklenburg: Yellow over red pennon; Corporals carried a yellow pennon with the Mecklenburg Arms thereon in red.
Oldenburg: Red over dark blue pennon; Corporals carried a dark blue pennon bearing a black heraldic Prussian eagle with a yellow star superimposed upon its breast.
Baden: Yellow over red pennon; Corporals carried a white pennon bearing a black griffon holding the gold/red/gold shield of Baden.
Hesse: White over red pennon; Corporals carried an all white pennon bearing the red and white striped Hessian lion.
Württemberg: Black over red pennon; Corporals carried a red pennon with the Württemberg Arms and crown in yellow thereon.

The 2nd. Prussian Guard Dragoon Regiment on parade. Note the Corporals' lance pennons.

Shabraque ornaments:

Left: The 1st. Prussian Guard Regiment on parade. Note the ornaments on the front and rear of the officer's shabraque.

PARADE SHABRAQUE *(Paradeuberlegedecke)*

The shabraques were carried only for parades and ceremonies and were, with the exception of the Hessian and Mecklenburg Regiments, cornflower blue cloth, fashioned with rounded corners both front and rear. The shabraque was trimmed with a broad band of facing colour cloth, set in from the outside edge, so that a strip of the cornflower blue cloth appeared on the outer side. The extreme outside edge itself had a further piping of facing colour, except for the 22nd. Regiment which had piping in red.

The Hessian Regiments Nos. 23 and 24 had a shabraque of similar pattern but in their case it was dark green, trimmed with the facing colour, and in the rear corners decorated with a white crown (silver for officers). The Hessian officers also had a plain blue service, or interim, shabraque.

The Mecklenburg Regiment No. 17 had a parade shabraque of black sheepskin for other ranks and black astrakhan for the officers, trimmed with poppy red cloth. The Mecklenburg Regiment No. 18 had a white sheepskin parade shabraque trimmed with black (velvet for officers). In both cases the front and rear corners were decorated with a Star surmounted by a Crown. This ornament was gold for officers and had the Arms of Mecklenburg superimposed on the Star in silver. The motif in the rear corner was larger than that on the front.

The Baden Regiments Nos. 20, 21, 22 had a red cloth Crown (silk for officers) on the rear of the shabraque.

Officers of the Prussian Guard Dragoon Regiments carried the Guard Star on both the front and rear corners of the shabraque. The N.C.O.'s and troopers had the Guard Star in the rear corners only.

Dragoons on manoeuvres showing saddlery, the position of the canteens and sabres in frogs.

BRIDLE AND SADDLERY *(Reitzeug/Sattel)*

The Prussian heavy cavalry bridle was of polished brown leather with steel buckles, tongues, rings, etc. The bit differed from the British universal pattern in having a thicker bridoon, and the bit had a considerably lower port and shorter cheek. A throat lash was included instead of a chain and the halter rein was usually coiled and looped to the lower part of it. The boss of the bit was circular and was decorated with the Royal cypher. The officers' bit differed in that the bridoon was omitted. The ends of the officers' straps were decorated with large silvered tips and their buckles were larger and hexagonal.

The Mecklenburg regiments had bridles similar to the hussar pattern with cross face piece with a brass plate over the union of the two straps, and a heart shaped plate on the martingale. The bit was 'S' shaped. The officers had gilded buckles and chain head band, etc. The Hessians had the Prussian type bridle but with the 'S' bit, and the two Wurttemberg regiments had Prussian pattern harness.

The saddle was lighter than the British pattern, higher in the arch, and with deeper and wider flaps. The girth protector flaps were nunmah and the panels were of the same material. A thick woollen blanket 7′ 7″ by 6′ 4″ was carried folded under the saddle. The stirrups were wider and heavier than the British counterpart and the girth was the so-called 'Cape' or Colonial fashion. A surcingle was used. The wallets were large and capacious. The nearside one contained cartridges and could have a shoe case strapped to the outside. A spade or hatchet could also be attached.

A short piece of rope with a ring at one end was carried over the carbine bucket; when a number of these ropes were joined together, a large picketing rope was fashioned which was

A dragoon orderly with carbine and lance.

The commanding officer of the 3rd. Baden Dragoon Regiment No.22 with undress bridle and saddlery.

passed over two long picketing staves which were usually carried in the regimental baggage.

A corn sack shaped like a sausage was strapped over the rear of the saddle and over the rolled greatcoat. A mess tin of aluminium, holding approximately 4½ pints, was carried in a leather case on the near side of the saddle. The tin cover could be used for a drinking utensil or as a frying pan and had a separate handle folded inside. The full weight carried by a troop horse in the field was approximately 100 lbs.

STANDARDS *(Standarten)*

In common with the other German cavalry, the standards of the dragoon regiments were decorated with various streamers commemorating campaigns, or a century or more of existence, or special parade streamers presented by royal personages.

The campaign streamers awarded, as applicable to the dragoons, were for the War of Liberation 1813/15, 1864 campaign against Denmark over the Schleswig-Holstein question, 1866 war against Austria-Hungary, and the 1870/71 war with France. Some regiments also had a special 1870 Iron Cross streamer awarded to those units which had already been granted the Iron Cross in the spear head of the standard haft, for services in the 1813/15 campaign. All the dragoon units also received in 1900 a special streamer presented by Kaiser Wilhelm II to all his regiments.

1st. Guard Dragoon Regiment: Swallow-tailed pattern, presented in 1893, the field being 'shot' white and silver material. The corner medallions and centre field of red and gold mixture material. Scroll white; embroidery, fringe, cyphers, etc. in gold.

Streamers: 1813/15, 1866, 1870/71, 1900, Parade streamer presented by Queen Victoria in 1893. Standard bearer's sash of light blue trimmed with silver lace.

2nd. Guard Dragoon Regiment: Square pattern, presented in 1861, with a field of silver material. The corner medallions the same as the field, the centre field orange. Scroll light blue, embroidery silver and green, silver fringe.

Streamers: 1866, 1870/71, 1900, Parade streamer presented by Empress Alexandra of Russia in 1914. Standard bearer's sash light blue trimmed with silver lace.

1st. Dragoon Regiment: Swallow-tailed pattern, presented 1890, having a black field with scarlet corner wedges. Scarlet corner medallions. White and silver mixture centre. Scroll black. Embroidery and fringe gold.

Streamers: 1813/15, 1866, 1870 Iron Cross, 1870/71, Centennial, 1900. Standard bearer's sash scarlet trimmed with gold lace.

2nd. Dragoon Regiment: Swallow-tailed pattern, presented in 1892, with a yellow field. Corner medallions yellow. Centre field white. Scroll yellow. Embroidery and fringe gold.

Streamers: 1813/15, 1866, 1870 Iron Cross, 1870/71, Double centennial, 1900. Standard bearer's sash black trimmed with gold lace.

3rd. Dragoon Regiment: Swallow-tailed pattern, presented in 1900, having a white field. Corner medallions white. Centre field light blue. Scroll white. Embroidery and fringe gold.

Streamers: 1813/15, 1866, 1870 Iron Cross, 1870/71, Double centennial, 1900. Parade streamer presented by King Gustav of Sweden in 1911. Standard bearer's sash rose-red with silver lace.

4th. Dragoon Regiment: Swallow-tailed pattern, presented in 1902, having a blue field with corners in gold material. Centre field 'shot' white and silver. Scroll blue. Embroidery and fringe gold.

Streamers: 1813/15, 1866, 1870/71, 1900. Standard bearer's sash light blue trimmed with silver lace.

Regiments 5 to 16 of the Prussian line regiments all carried the pre-1888 model standard. This was the square pattern and had a black field with white corner wedges, the centre field being orange and the motto scroll light blue. Embroidery of the monograms and crowns in gold and the fringe silver.

The streamers applicable to the various regiments were as follows:-

5th. Regiment	Standard presented 1861: 1866, 1870/71, 1900.
6th. Regiment	Standard presented 1861: Streamers as for Regiment 5.
7th. Regiment	Standard presented 1861: Streamers as for Regiment 5 but with the addition of the 1864 streamer with swords.
8th. Regiment	Standard presented 1861: Streamers as for Regiment 5.
9th. Regiment	Standard presented 1867: 1870/71, 1900.

Regiments 10 to 16 all received their standards in 1867 and all carried the same streamers as the 9th. Regiment but Regiment 12 had, in addition, a parade streamer presented by Princess Eitel Friederich in 1910. In each case the standard bearer's sash was faced with cloth of the distinctive colour and trimmed with lace of the button colour.

17th. (Mecklenburg) Dragoon Regiment: Square pattern, with a white field. In the centre the Arms of Mecklenburg surmounted by a crown. In each corner a medallion with the cypher FF surrounded by a wreath of palm and laurel leaves, all in gold embroidery. Gold fringe. Staff white.

Streamers: Prussian 1866, Ribbon and cross of the Mecklenburg Militär-Verdienstkreuz for 1870/71, special parade streamer presented by Grand Duchess Anastasia in 1881, 1900. Standard bearer's sash light blue trimmed with gold lace.

18th. (Mecklenburg) Dragoon Regiment: Also of square pattern with a white field. Again, the centre bore the Arms of Mecklenburg and supporters. In each corner the cypher FF surmounted by a crown and surrounded by a wreath of palm and laurel leaves, all in gold embroidery, the corner medallions being linked by an ornamental gold border. Fringe gold. Staff white.

Streamers: Ribbon and cross of the Mecklenburg Militär-Verdienstkreuz for 1870/71, Prussian 1870/71, 1900, the old Mecklenburg banderole (gold with black thread stripes). Standard bearer's sash light blue trimmed with gold lace.

The standard bearer's gorget for the Mecklenburg dragoons was the same pattern as the Prussian except that it carried the cypher FF thereon. However, the arm badge was completely different, this being a metal oval wreath surrounding the cypher FF, the whole topped by a crown and all in the button colour.

19th. (Oldenburg) Dragoon Regiment: Square pattern with a light blue field in the centre of which was the Arms of Oldenburg, surmounted by a crown, the whole surrounded by a wreath of laurel in natural colours. In the corners the cypher FFA surrounded by a laurel wreath in natural colours and surmounted by a crown. All the embroidery and fringe in gold. The staff was dark blue.

Streamers: Oldenburg commemorative medal and ribbon for 1866, Prussian 1866, 1870/71, 1900, the old Oldenburg banderole (blue, edged red). Standard bearer's sash light blue trimmed with silver lace.

20th. and 21st. (Baden) Dragoon Regiments: Square pattern, white field. In the centre the Griffon of Baden holding a shield and sword in gold. The whole surrounded by a laurel

The standard of the Baden Dragoon Regiment No.20.

wreath in green. Above the Griffon a red scroll with the motto "Fidelitas" in gold. Overall a gold crown with a red lining. In each corner a gold cypher comprising an intertwined double F surmounted by a gold crown with red lining and surrounded by a green laurel wreath. The fringe gold. Staff black.

Streamers: Silver Militär-Verdienst medal on its ribbon, 1900, the old Baden banderole (silver with two red/gold stripes). The spearpoint of the haft was decorated with the Iron Cross for the Franco-Prussian campaign of 1870/71. Standard bearer's sash light blue with silver lace.

22nd. (Baden) Dragoon Regiment: As for the 20th. Regiment but with the addition of the Prussian streamer for 1870/71 (awarded in 1895).

23rd. (Hessian) Regiment: Square pattern. Field of silver material. In the centre a blue shield bearing the red and white striped Hessian Lion rampant, supported by two golden lions and above the shield a gold crown. Behind the shield a trophy of arms and beneath it the Cross of the Order of Ludwig suspended from a black and red striped ribbon. In each corner a gold

crowned cypher L II surrounded by a wreath of laurel and oak leaves in green. Gold fringe. Staff black and gold.
Streamers: A cravat in the colours of the Ludwig Order, Cravat and cross of the Militar-Verdienstkreuz, a field service cravat, Centennial (1790), 1900, Parade streamer presented by Princess Heinrich of Prussia in 1890, the old Hessian banderole (silver and red). The spearpoint of the haft was decorated with the Iron Cross for the Franco-Prussian campaign. Standard bearer's sash red and silver.
24th. (Hessian) Dragoon Regiment: Square pattern. Field of silver material with light blue corners. In the centre the coat of arms of the Landgrave of Hesse in full colours, surrounded by trophies of war. Underneath, two drums and the badge of the Order of the Black Eagle. In each corner an intertwined cypher L surrounded by a wreath and topped with a crown, all in gold. The lining of the crowns in red. Gold fringe. Staff dark blue and gold.
Streamers: Honour medallion of Ludwig I, Cravat and cross of the Militar-Verdienstkreuz, 1900, Parade streamer presented by Czar Nicholas II of Russia in 1896, the old Hessian banderole. The spearpoint of the haft was decorated with the Iron Cross for the 1870/71 campaign. Standard bearer's sash red with silver lace.
25th. (Württemberg) Dragoon Regiment: Square 'Vexillum' pattern. Field red. On the front the Arms of Württemberg supported by the black lion of Swabia and the golden hart of Württemberg. Beneath it a blue scroll bearing the inscription *"Furchtlos und Treu"* in silver, together with the white cross of the Militär-Verdienst Order suspended in the middle. On the reverse side of the standard a red and gold crowned W. Fringe yellow. Staff black.
Streamers: Prussian 1870/71, Centennial streamer (presented 1913), 1900, Parade streamer presented by Kaiser Wilhelm II in 1913.
26th. (Württemberg) Dragoon Regiment: The 26th. Dragoons had two standards. The regimental standard followed the pattern of that of the 25th. but had green fringes.
Streamers: Centennial (awarded in 1905), and 1900.

The regiment possessed in addition a special honour standard awarded for bravery at the Battle of Linz on the 17th May, 1809. It was also of square 'Vexillum' pattern and had a green field (the regiment's old colours). In the centre, the badge of the Militar-Verdienst Order in gold with a laurel wreath in green below; the whole surrounded by a gold crown. In each corner a white medallion containing the cypher FR in gold surmounted by a gold crown. On the rear side of the standard a gold cypher FR topped by a gold crown. Fringe gold. The top of the staff was surmounted by a gold eagle in place of the normal spear point, and the staff itself was black.
Streamers: Special parade streamer and 1900. Standard bearer's sash light blue with silver lace. The standard bearer's gorget and arm badge bore the cypher "W" and the three harts.

1900 Streamers

As mentioned above, Kaiser Wilhelm presented to all his regiments at the beginning of 1900 a special streamer to mark the advent of the new century. In the case of the Prussian regiments the streamer was silver with three black stripes and bore two gilt clasps. One clasp had on one side the Imperial crown and on the other the date *1. Januar 1900*; the other bore on one side the crowned monogram W (Wilhelm II) and on the reverse two dates, *1 Januar 1900* and, below it, the foundation date of the regiment.

In the case of the Contingent states the streamers were in the state colours and the clasp carried the monogram of the ruler instead of the W cypher.

The colours and clasps were as follows, for the dragoon regiments:-

17th. and 18th. Mecklenburg Dragoons: Streamer: Red with light blue-yellow-light blue narrow stripes at each edge. Clasp monogram: LPR (Luitpold Prinz Regent).

19th. Oldenburg Dragoons: Streamer: Half blue and half red. Clasp monogram: P (Peter).

20th., 21st. and 22nd. Baden Dragoons: Streamer: Yellow with a red central stripe and narrow white edges. Clasp monogram: F (Friedrich I).

23rd. and 24th. Hessian Dragoons: Streamer: Silver with three red stripes. Clasp monogram: EL (Ernst Ludwig).

25th. and 26th. Württemberg Dragoons: Streamer: Dark red, a black stripe at each edge. Clasp monogram: W (Wilhelm II).

A dragoon regiment on parade. The regimental colours are covered. The officer is using the undress saddlecloth.

KETTLE DRUM BANNERS *(Paukenfahnen)*

By 1914, ten regiments of dragoons possessed kettle drums, mostly as a result of gifts from their princely Colonels-in-Chief. They were as follows:

Regiment	**Origin of drums**
1st. Guard Dragoon Regiment	Presented in 1894 by Queen Victoria.
2nd. Guard Dragoon Regiment	Presented in 1896 by the Empress Alexandra of Russia.
1st. Dragoon Regiment	Carried off as spoils of war after capturing them from the Saxon Guard Carabiniers at Kesseldorf in 1745.
2nd. Dragoon Regiment	Carried after capturing them in 1810 from the Bavarian Dragoon Regiment "Konig von Bayern".
3rd. Dragoon Regiment	Presented in 1885.
9th. Dragoon Regiment	Presented in 1907 by King Carol I of Rumania.
17th. (Mecklenburg) Dragoon Regiment	Presented in 1894 by Grand Duke Friedrich Franz III.
20th. (Baden) Leib Dragoon Regiment	Presented with copper drums in 1878 by Grand Duke Friedrich I of Baden, and changed to silver in 1879.
23rd. (Hessian) Guard Dragoon Regiment	Presented in 1877 by Grand Duke Ludwig IV.
24th. (Hessian) Leib Dragoon Regiment	Presented in 1903 by Czar Nicholas II of Russia.

Each pair of kettle drums had two sets of banners; a set used for all normal occasions and a set kept specially for parades and ceremonials. The normal duty banners *(interimspauken-fahnen)* were, with the exception of the Hessian Dragoon Regiment No. 24, universally of cornflower blue cloth with trimmings and tassels in wool of the button colour. The parade banners were much more elaborate and were made from velvet and silk with gold or silver lace and embroidery.

There were two variations in the pattern of the parade drum banners. The older type distributed before 1888 (i.e. to Dragoon Regiments 1, 2 and 3) had panels cut with rounded edges and the upper panels only partly covered the lower, so that both panels were equally visible; both upper and lower panels had fringing and embroidery. A new pattern was introduced in 1888, however, in which the panels were cut to a point so that the upper panels were very long and almost completely covered the lower. For that reason, the upper panels only had embroidery and emblems. Both rows, however, had fringing and tassels. The colours of the parade banners and the regiments entitled to carry them are shown below:-

1st. Guard Dragoon Regiment: The upper panels of blue velvet, the lower panels of red silk. Lace and embroidery gold. The panels carried the following designs or emblems; the cypher WR, the Guard Star, the Arms of Great Britain, the Prussian heraldic eagle, the cypher VIR, the date 21.2.1894 encircled by a wreath.

A Trumpeter of Dragoon Regiment No.25 in parade dress.

A Vizewachtmeister of Dragoon Regiment No.23 in parade dress.

An Adjudant of Dragoon Regiment No.8 in undress uniform.

A Major of Dragoon Regiment No.7 in walking out dress and überrock.

A Lieutenant of Dragoon Regiment No.24 in walking out dress and überrock.

A Gefreite of the Grenadier Regiment zu Pferde. Note the hand grenade ornament on the chin scale boss.

A Major of the 2nd. Guard Ulan Regiment in service dress.

An officer of Dragoon Regiment No.13 in umhang and manöverkappe.

A Wachtmeister of Dragoon Regiment No.2 wearing the litewka.

A One-Year Volunteer Gefreite of Dragoon Regiment No.24.

Dragoon and Chevauleger undress caps.
Top row: 1st. Guard Dragoon Regiment; 2nd. Guard Dragoon Regiment; Dragoon Regiments Nos. 1, 2, 3, 4. *2nd. row:* Dragoon Regiments Nos. 5, 6, 7, 8, 9, 10. *3rd. row:* Dragoon Regiments Nos. 11, 12, 13, 14, 15, 16. *4th. row:* Dragoon Regiments Nos. 17, 18, 19, 20, 21, 22. *5th. row:* Dragoon Regiments Nos. 23, 24, 25, 26; Chevaulegers Regiments Nos. 1, 2. *6th. row:* Chevaulegers Regiments Nos. 3, 4, 5, 6, 7, 8.

Above: The standard of the 1st. Guard Dragoon Regiment.

Below: The standard of the 2nd. Guard Dragoon Regiment.

Above: The standard of Dragoon Regiment No.19.

Below: The standard of Dragoon Regiment No.20.

The parade drum banners of Dragoon Regiment No.9

The standard of Dragoon Regiment No.12.

The standard of Dragoon Regiment No.17.

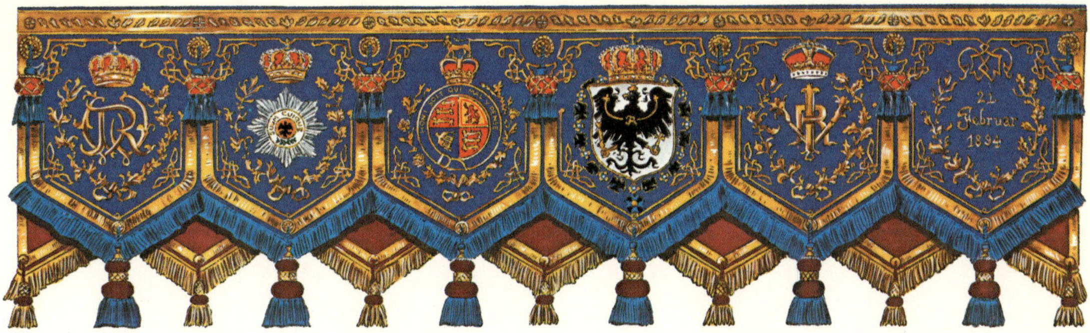

The parade drum banners of the 1st. Guard Dragoon Regiment.

The standard of Dragoon Regiment No.23.

The standard of Dragoon Regiment No.24.

Above: An Oberst Leutnant of the 1st. Guard Ulan Regiment in parade dress.

Below: A trooper of the 2nd. Guard Ulan Regiment in parade dress.

Above: The Musikmeister of the 3rd. Guard Ulan Regiment in parade dress.

Below: The Adjudant of Ulan Regiment No.9.

Above: The Regimental Adjudant of the 1st. Bavarian Ulan Regiment in parade dress.

Below: A One-Year Volunteer Unteroffizier in walking out dress.

Above: A junior officer of the 2nd. Bavarian Ulans in service dress.

Below: The Regimental Adjudant of Ulan Regiment No.21 in undress uniform.

Ulan undress caps.
Top row: 1st. Guard Ulan Regiment; 2nd. Guard Ulan Regiment; 3rd. Guard Ulan Regiment; Ulan Regiments Nos. 1, 2, 3. *2nd. row:* Ulan Regiments Nos. 4, 5, 6, 7, 8, 9. *3rd. row:* Ulan Regiments Nos. 10, 11, 12, 13, 14, 15. *4th. row:* 16, 17, 18, 19, 20, 21. *5th. row:* 1st. Bavarian Ulan Regiment; 2nd. Bavarian Ulan Regiment.

The standard of Ulan Regiment No.3.

The parade drum banners of Ulan Regiment No.5.

Lance pennons.

The standard of Ulan Regiment No. 13.

The standard of Ulan Regiment No.7.

The standard of the 1st. Bavarian Ulans.

The standard of Chevauleger Regiment No.1.

An officer of Chevauleger Regiment No.3 in parade dress.

A Gefreite of Chevauleger Regiment No.5 in service dress.

2nd. Guard Dragoon Regiment: The upper panels of orange velvet, the lower of red silk. Lace and embroidery silver. The panels carried the following emblems; the cypher WR, the Prussian heraldic eagle, the Royal Arms of Russia, the Royal Arms of Hesse, the cypher AI.
1st. Dragoon Regiment: Red velvet over red silk. Gold and silver lace and embroidery. Each of the upper panels bore the Prussian heraldic eagle on a silver shield surrounded by a wreath of laurel and palm leaves. In each of the lower panels a crowned WR cypher in gold within a wreath of laurel and palm leaves.
2nd. Dragoon Regiment: Black velvet over white silk. Gold and silver lace and embroidery. Emblems and motifs as for 1st. Regiment.
3rd. Dragoon Regiment (Grenadiers zu Pferde): Rose pink velvet over blue silk. Gold and silver lace and embroidery. Emblems and motifs as for the 1st. Regiment.
9th. Dragoon Regiment: Blue velvet over black velvet. Gold and silver lace and embroidery. Fringing of light blue and silver mixture. Motifs and emblems; the Prussian Arms, cypher CI, Rumanian Arms, cypher WRII, the Prussian Arms.
17th. Dragoon Regiment: Red velvet over cornflower blue silk. Lace and embroidery gold. Motifs and emblems: the Arms of the Grand Duchess Maud of Mecklenburg and Anastasia of Russia, painted thereon. The cornflower blue lower panels bearing the letters M & P embroidered in gold.
20th. Dragoon Regiment: All red velvet, with fringing, lace, tassels, etc., in silver. Alternating emblems consisting of intertwined letters L & F (Luise and Friedrich), embroidered in silver and surrounded by a silver laurel wreath, surmounted by a gold crown with red linings, and the Arms of Baden in gold/red/gold surmounted by a gold crown with red linings, all within a silver laurel wreath.
24th. Dragoon Regiment: This was a completely different pattern to any of the foregoing drum banners. There was only one row of panels in dark green cloth, with each panel flanked by a hanging embroidered ribbon which was loose. The panels bore the emblems: Royal Arms of Russia, cyphers NII and EL, Royal Arms of Hesse. Embroidery, etc., in silver.

TRUMPETS *(Trompeten)*

The signal instrument of the cavalry was the trumpet. The traditional instrument was the so-called fanfare trumpet, first introduced in the 18th. century. It was a long brass trumpet with a double bend and was engraved around the bell with baroque decorations. After 1817, a shorter model was taken into use which, with few modifications, continued in use until 1914. This instrument had one bend and was approximately half the length of the old *fanfaren-trompete.* An order of December, 1895, stated that the trumpet was to be 42 cm. long.

The normal trumpet cords were in state colour and white wool with large N.C.O. pattern tassels, gold or silver cords being used by some regiments for parades. The Trumpet-Major had cords of state colour mixed with silver. The Mecklenburg regiments were an exception; their trumpet cords were yellow, mixed with light blue and red.

If the old fanfare trumpet was carried in parade, it had special netted cords with four additional tassels hanging from the length of the double bend. The 1st. Regiment had silver trumpets from 1721.

The regimental mounted band of the 1st. Prussian Guard Dragoon regiment.

HORSE COLOURS FOR TRUMPETERS AND KETTLE DRUMMERS

Regiment	Trumpeters	Kettledrum horse where known
1st. Guard	Chestnuts	Chestnut
2nd. Guard	Blacks	Black
1st.	Chestnuts	Dappled grey
2nd.	Blacks	Grey
3rd.	Chestnuts	Chestnut
4th.	Chestnuts	
5th.	Mixed: Chestnuts, Browns and Greys	
6th.	Blacks	
7th.	Chestnuts	
8th.	Blacks	
9th.	Bays	Piebald
10th.	Blacks	
11th.	Chestnuts	
12th.	Chestnuts	
13th.	Mixed: Chestnuts, Bays and Blacks	
14th.	Blacks	
15th.	Blacks	
16th.	Blacks	
19th.	Blacks	
20th.	Chestnuts	Chestnut with dark feet and mane
21st.	Blacks	
22nd.	Brown	

SPECIALIST INSIGNIA *(Sonderheiten Abzeichen)*

Fencing distinctions *(Fechterabzeichen):* These badges were first authorised for the cavalry in 1889 and comprised a series of chevrons worn on the right upper arm. The standard-bearers' shield and the signallers' badges were worn above them. For dragoons the chevrons were as follows:-

1st. Class	1 woollen chevron in the state colours
2nd. Class	2 woollen chevrons in the state colours
3rd. Class	3 woollen chevrons in the state colours
4th. Class	1 gold or silver lace chevron
5th. Class	1 gold or silver lace chevron and one woollen chevron
6th. Class	1 gold or silver lace chevron and two woollen chevrons
7th. Class	1 gold or silver lace chevron and three woollen chevrons
8th. Class	2 gold or silver lace chevrons
9th. Class	2 gold or silver lace chevrons and one woollen chevron
10th. Class	2 gold or silver lace chevrons and two woollen chevrons
11th. Class	2 gold or silver lace chevrons and three woollen chevrons
12th. Class	3 gold or silver lace chevrons

A group of senior N.C.O.'s showing a variety of specialist insignia, including farriers badge, sharpshooting distinction and fencing distinction.

Specialist insignia:

Flaggenwinkerabzeichen

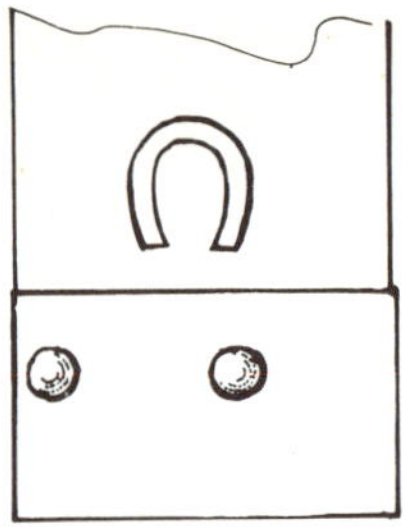

Beschlagschmied

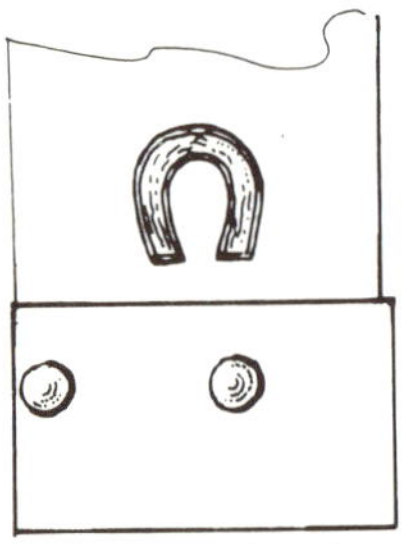

Fahnenschmied

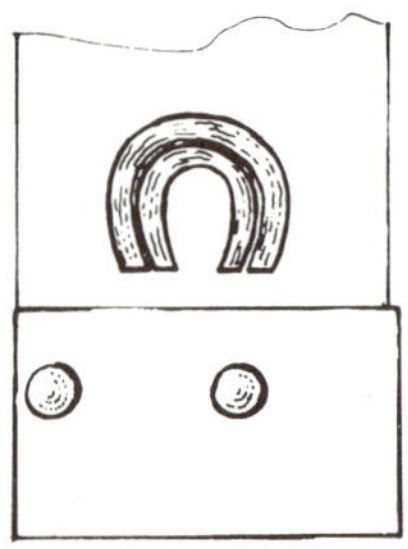

Oberfahnenschmied

Right: A Sergeant of Dragoon Regiment No.25 showing the schützenabzeichen.

Sharpshooters' badges *(Schutzenabzeichen):* From 1894 the cavalry took into use the infantry pattern sharpshooters' badges. These comprised eight classes of plaited lanyards with rosettes and acorns and were worn on the left side of the full dress tunic suspended from the shoulder strap to the second button from the top. The cord was a mixture of black, white, red woollen cord for the first three classes with the addition of one, two or three acorns. The fourth class had the lanyard in silver and silk cord with no acorn and the fifth class the same plus a yellow metal crowned medallion. The sixth to eighth classes wore the same cord but with the addition of one, two or three acorns respectively. The acorns were worn at the button end and the medallions at the shoulder. For the Dragoon the cord was worn on the left shoulder and for the Ulan on the right shoulder.

Farriers *(Schmiede):* The farrier *(Beschlagschmied)* wore the regimental uniform with a small cloth horseshoe in the facing colour above the cuff of the tunic. The Corporal-Farrier *(Fahnenschmied)* wore a similar badge but in gold or silver lace according to button colour, and the Sergeant-Farrier *(Oberbeschlagsschmied)* wore the same badge but with a larger gold or silver horseshoe enclosing the smaller one. The N.C.O.'s wore the other distinctions of their rank where appropriate. Farrier N.C.O.'s and smiths were under the command of the Veterinary Staff.

Riding School Badges *(Reitschulenabzeichen):* Personnel who had completed a one year course at the Military Riding Institute at Hanover wore a twisted cord in the button colour

Specialist insignia:

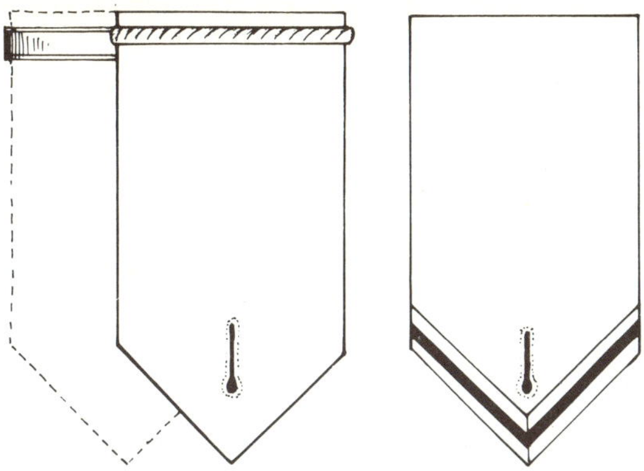

Reitschulenabzeichen Hanover   Militar-telegraphen abzeichen

Right: The Regimental Adjudant of the 1st. Prussian Guard Dragoon Regiment. He is wearing the sash with the undress cap, which was contrary to dress regulations. Note the absence of the waist-belt.

across the sleeve end of the shoulder strap. The special badge for two courses was a flat worsted strip in the button colour with corded edges worn in the same position. Saxons wore a special badge comprising a button colour metallic strip above the cuff until 1905.

Telegraphists' Badge *(Militär-Telegraphen Abzeichen):* A strip of braid in the state colours was worn along the edge of the point end of the shoulder strap ('A' shaped). The Bavarian Regiments had an additional similar coloured strip down the centre of the shoulder strap.

Signallers' badges *(Flaggenwinker-Abzeichen):* A circular piece of cloth of the tunic colour with two crossed flags with yellow staves. The rear flag white, the front flag red. This badge was worn on the right upper sleeve unless the recipient had *fechterabzeichen* and *flaggenwinkenabzeichen* in which case it was worn on the left lower sleeve.

REGIMENTAL ADJUTANTS

Regimental adjutants wore a sash of silver or gold thread with a stripe of the state colours towards either edge over the right shoulder. These special sashes were first introduced into the Prussian Army in 1848, the other German States following suit shortly after. The sash was secured by a bridle fastening above the left hip and had two large fringed tassels. The pouch belt, waist sash, and waistbelt were not worn by Regimental Adjutants.

Shoulder strap of a One-Year Volunteer. The colours of the cord were as follows: Prussia—black and white; Bavaria — light blue and white; Saxony — green and white; Württemberg — 1 red, 1 black and 4 white; Hesse — 1 red, 1 white and 1 red; Mecklenburg — red, yellow, blue, red, yellow, blue.

ONE-YEAR VOLUNTEERS *(Einjährig-Freiwillige)*

The one year volunteers were young men, usually of solid middle class background who, having attained an advanced standard of academic qualifications, were privileged to serve only one year with the Colours instead of the obligatory three of their less fortunate comrades. The One Year Volunteer had a number of privileges; for example, he was permitted to choose his regiment and was allowed to live out. He shared the service of a batman along with several other one-year volunteers and usually messed with other volunteers in a restaurant. Generally they could rise to the rank of sergeant and returned to serve their subsequent reserve service as vizewachtmeister until they qualified for a commission. As Reserve Officers they usually rose to the rank of Captain but were generally speaking treated as inferior to the regular officer class.

The distinguishing badge of the dragoon one-year volunteer was a twisted cord in the state colours around the shoulder strap. This twisted cord known as the *schnur* was white with black for the Prussians, white with light blue for the Bavarians, white with green for the Saxons, red, black and white for the Württemberg Regiments, yellow, red and cornflower blue for Mecklenburg, and red and white for Hesse.

THE STANDARD BEARER *(Standartenträger)*

The standard bearer was usually a sergeant and chosen from one of the best swordsmen holding that rank. The badge of the appointment was a shield, in cloth of the tunic colour, worn on the right upper sleeve above the swordsmanship chevrons. In addition, the standard-bearer wore a gorget suspended on a chain around his neck. The gorget was in brass or German silver, according to the button colour, and had in relief of the opposite metal devices comprising the crossed flags and a central cypher, crown or badge, and grenades in the corners.

The Prussian gorget badge comprised crossed flags with the WRII cypher and crown. The arm badge consisted of crossed flags with a crown between the flags and WII beneath. The Guard regiments had similar gorgets but the central cypher was replaced by the Guard Star under the crown. The arm badge for Guard regiments was the same as for the line cavalry.

Baden and Oldenburg regiments wore the same gorget and arm badge as the Prussians. Mecklenburg had an FRIII cypher in the centre of the gorget, with a crown over. The arm badge was circular and consisted of a blue cloth badge with an oakleaf wreath, central cypher and crown, all in fret cut metal. The Hessians had an EL cypher and crown on the gorget and an arm badge with a Hessian cypher in lieu of the Prussian WII. The Württemberg regiments had Prussian style gorgets and arm badges, but with a Gothic W and crown on both gorget and arm badge.

The standard was supported on a red leather bandolier, with a double carbine hook at the bottom end. The bandolier was faced with cornflower blue cloth and had lace in the button colour along either edge, leaving a piping of blue on the outside edge. The bandolier of the 1st. Guard Dragoons was fringed, and in the facing colour. The bandoliers of the 1st., 2nd., and 3rd. Regiments were also in the facing colour.

Gorgets and arm badges of the standard bearers of the various states with a detail of the grenade ornamentation.

A dragoon standard bearer, taken from a sketch by an Italian artist dated 1909.

THE TRUMPET-MAJOR/BANDMASTER
(Stabstrompeter/Musikmeister)

Before 1908, the Trumpet-Major, who led the band, wore the uniform and rank insignia of a Sergeant-Major. He was distinguished by wearing the red plume of a trumpeter on his helmet and bandsmens' wings which had, additionally, long fringes in gold or silver, according to the button colour. He also wore special shoulder ornaments of interlaced cord. If the Trumpet-Major attained the honorary title of Conductor of Music *(Musik-Dirigent),* a narrow braid of metallic lace was added to the centre of the shoulder cords; if he was further promoted to Director of Music *(Musik-Direktor),* a further metallic braid was added to the cords.

In 1908 the status of these categories was enhanced by the creation of the rank of Bandmaster, who although still counted as N.C.O.'s, had a rank senior to the Sergeant-Major and received a distinctive and quasi-officer type uniform. The collar of the tunic was cut square, was higher than the other N.C.O.'s, and was hooked tight. Their rank braid was in white or yellow silk or camel hair. An officers' pattern greatcoat was taken into use, but the underside of the collar was grey when turned up instead of the regimental facing colour. Horse furniture continued to be as for the other ranks but they had officers' sword-knots. Belts, etc., were the same pattern as the other ranks but were usually lacquered.

The uniform was further distinguished by shoulder cords. These were now larger and generally poppy red, except for the 3rd., 7th., and 15th. Regiments, which had rose red, and the 11th. and 12th. Regiments, which had crimson cords. The cords were further decorated with gilt lyres. Bandmasters of regiments with devices on their shoulder straps wore these in addition to the lyres. After about five years in the rank it was possible for the holder to be promoted to Senior Bandmaster *(Obermusikmeister)* and this was recognised by the addition of a thin golden thread to the centre of the cords.

Bandmasters continued to wear the fringed bandsmens' wings but the lace was now silver and instead of the bars being diagonal they were cut in points and were vertical. At each edge of the lace was a thin thread in the state colours and at the base of the vertical laces was a horizontal silver lace below which the fringing was suspended. Bandmasters' trumpet cords were in silver or gold mixed with the state colours.

2: Lancers

As early as 1704, the Prussians were using cavalry troops armed with lances. In 1745, a unit known as "Bosniaken" was linked with the "Totenkopf-Husaren-Regiment" and wore a costume which derived from Bosnian and Polish national costume.

By 1806, the Prussian Army still retained units with strong Polish/Bosnian links. These were the "Towarczys" Regiment and a self-standing "Towarczy" Battalion, both of which were armed primarily with the lance, and were used for scouting and foraging duties for which the Ulans became justly famous. These troops were used in East Prussia and the bordering Polish and Austrian-Hungarian states and had a distinctly Tartar-Slav appearance.

Saxony also had its 'Ulanen-pulks' in the 18th. century, and in Bavaria a lancer regiment was raised in 1813, dressed in the 'Österreichischer' style.

The Prussian Ulan regiment, raised in 1806/7 from the former "Towarczys", had a strength of eight squadrons, but was later split in two to form two separate regiments of four squadrons each. By 1819, the establishment of Ulan Regiments had been increased to two Guard Regiments and eight line regiments distinguished by facing colours and different coloured buttons.

The Garde-Regiments stemmed from the Leib-Ulanen-Eskadron re-titled the Garde-Ulanen-Eskadron in 1810; a second Garde-Ulanen-Regiment was raised in 1819. The line regiments were raised from the Freiwillige and National cavalry regiments of the 1813-1815 Freedom War, the Freikorps Lützow, Freikorps Hellwig and the Russo-German Legion. In 1860, besides a 3rd Garde-Regiment, a further four line regiments were formed. In 1856, the third line regiment received the title Ulanen-Regiment König Alexander II von Russland. In 1866, Regiments 13-16 were formed. In 1884 the "West Preussen Ulanen Regiment No.1" was titled "Ulanen-Regiment Konig Alexander III von Russland (West Preussisches) No.1". In 1889 Kaiser Wilhelm II was declared the Regimental Chief of the "1st Hannoverschen Ulanen-Regiment No. 13" which, on the same day, was re-titled "Konigs-Ulanen-Regiment (1 Hannoversches) No. 13". In 1902 the "Ulanen-Regiment Hennings von Treffenfeld (Altmarkisches) No. 16", was granted the distinction of wearing the cypher of King George of Saxony, and in the next year the "Thüringische-Ulanen-Regiment No. 6", was granted similar distinction in respect of King Christian IX of Denmark.

By 1871 the Württemberg cavalry had been re-organised as two Dragoon and two Ulan Regiments. The Ulans became the 19th. and 20th. Ulan Regiments in the national numbered sequence. The Royal Armies of both Saxony and Württemberg retained their own War Ministries and Headquarters Staffs and Establishments and were bracketed in the 1st. and 2nd. Royal Saxon Army Corps (XII and XIX of the National Army) and the Royal Württemberg Army Corps (XIII of the National Army). These armies retained many characteristics of their old uniforms. The 1st. and 2nd. Saxon Ulans which were both raised in 1867 were given the numbers 17 and 18 in the national numbered sequence in 1871. A further Saxon Regiment was raised in 1905 and took the number 21.

The Royal Bavarian Army remained completely autonomous under the command of its King and with its own Headquarters Staff and Establishments. It was formed in the I, II and III Bavarian Army Corps and retained its old uniforms but accepting some characteristics of the Prussian uniform and adapting these to their own needs. Both the 1st. and 2nd. Bavarian Ulan Regiments were raised on the 21.12.1863 and did not figure in the numbered sequence of the national army.

GENERAL DEVELOPMENT OF THE ULAN UNIFORM

The original uniform coat of the line regiments was the Kollet, a short waisted double breasted coat with red collar, cuffs and short turn-back skirts. The regiments were distinguished by the colour of the shoulder strap. The headdress was a leather and felt chako with a large woollen cockade on the front and long cap lines "fangschnur" which looped around the body after passing under the right shoulder strap. The Leib-Eskadron wore a uniform coat which was in the so-called *Österreichischen* style with narrow red lapels and white woollen fringed epaulettes. Later the Garde-Ulanen-Eskadron also took the kollet into use, but with epaulettes comprising metal crescents and chain reinforcement along the straps. The 4th. Regiment received the uniform of the earlier regiments but with light blue shoulder straps and Regiments 5 to 8 were the same but with white metal buttons.

By 1824 all Ulan regiments had taken into use the crescent-ended shoulder ornaments and by 1843 the kollet with the traditional plastron shaped lapels had appeared. The ulan tunic known as the ulanka was authorised in 1853.

The chako had been discarded by 1816 and replaced by a Polish style headdress with a rectangular flat top known as the tschapska. This was now used by all ulan regiments.

The style of the headdress altered with the fashion of the period becoming lower in form and with a smaller flat top until by 1867 it had achieved the classic shape which remained substantially unaltered until it reached its ultimate form in 1889.

Regiment	Shoulder Straps	Buttons
'Garde-Ulanen-Regiment'	Red epaulettes	Yellow
1 (Westpreussisches)	White	Yellow
2 (Schlesisches)	Red	Yellow
3 (Brandenburgisches)	Yellow	Yellow
4 (Pommersches)	Light Blue	Yellow
5 (Westfalisches)	White	White
6 (2 Westfalisches)	Red	White
7 (1 Rheinisches)	Yellow	White
8 (2 Rheinisches)	Light Blue	White

By 1851 the 1st. Guard Regiment had white Garde lace and white buttons, and the 2nd. Guard Regiment, yellow (with a red centre 'light' in the lace in both cases). The 9th. to

12th. Regiments, created in 1860, departed from the original principle of red facings and took into use the following colours for the collar, cuffs, plastron, pipings, epaulette field and girdle trimming: 9th.—white; 10th.—crimson; 11th.—yellow; 12th.—light blue. Buttons were all yellow as were the metal crescents of the epaulettes. The 3rd. Guard Regiment raised at the same time had yellow facings and white Garde lace and metal, but with yellow metal crescents.

Regiments 13 to 16 were formed in 1866 and repeated the regimental distinctions of Regiments 9 to 12, that is to say, white, crimson, yellow and light blue respectively, except in this case, the buttons were white. However the epaulette crescents remained yellow for all regiments except the 15th. who had the distinction of having German Silver epaulette crescents.

In 1861, the cuffs of the ulanka were made smaller and by 1868 the collar was made lower. In 1889, white tunic piping was introduced for Regiments 12 and 16 and in 1894 this was extended to the top edge of their cuffs. Their plastrons and girdle trimming were also edged white.

By an order of the 24.1.1899, German Silver epaulettes were introduced for the 13th. Regiment, and a further order in 1907 decreed the same for the 1st. and 3rd. Guard Regiments to replace the brass types then in use.

ULANS 1900/12

Title	Raised	Garrison	Army Corps
Garde Ulanen Regiment No.1	14. 4.1819	Potsdam	Garde
Garde Ulanen Regiment No.2	14. 4.1819	Berlin	Garde
Garde Ulanen Regiment No.3	7. 5.1860	Potsdam	Garde
Ulanen Regiment Kaiser Alexander III von Russland (West Preussisches) No.1	1. 8.1745	Militsch, Ostrowo	V
Ulanen Regiment von Katzler (Schlesisches) No.2	15. 5.1809	Gleiwitz, Pless	VI
Ulanen Regiment Tsar Alexander II von Russland (Brandenburgisches) No.3	15. 5.1809	Fürstenwalde	III
Ulanen Regiment von Schmidt (Pommersches) No.4	7. 3.1815	Thorn	XX
Ulanen Regiment (Westfälisches) No.5	7. 3.1815	Düsseldorf	VII
Ulanen Regiment (Thüringisches) No.6	18. 2.1813	Mülhausen, Hanau, Langensalza	XVIII
Ulanen Regiment (Grossherzog Friedrich von Baden) (Rheinisches) No.7	14. 4.1734	Saarbrücken	XXI
Ulanen Regiment Graf zu Dohna (Ostpreussisches) No.8	6. 8.1812	Stallupönen, Lyck	I
Ulanen Regiment (2 Pommersches) No.9	7. 5.1860	Demmin	II
Ulanen Regiment Prinz August von Wurttemberg (Posen) No.10	7. 5.1860	Züllichau	V

Title	Raised	Garrison	Army Corps
Ulanen Regiment Graf Haeseler (2 Brandenburgisches) No.11	7. 5.1860	Saarburg	XXI
Ulanen Regiment (Litthauisches) No.12	7. 5.1860	Insterburg, Stallupönen, Goldknap	I
Ulanen Regiment Königs (Hannoversches) No.13	19. 2.1813	Hanover	X
Ulanen Regiment (2 Hannoversches) No.14	10.12.1805	St. Arvold, Marchingen	XVI
Ulanen Regiment (Schleswig-Holstein) No.15	27. 9.1866	Saarburg	XXI
Ulanen Regiment Hennig v Treffenfeld (Altmärkisches) No.16	27. 9.1866	Salzwedel & Gardelegen	IV
Ulanen Regiment (Kaiser Franz Josef von Österreich, König von Ungarn) No.17	1. 4.1867	G. Oschatz	XII
Ulanen Regiment (2 Saxon) No.18	1. 4.1867	Leipzig	XIX
Ulanen Regiment (König Karl) (1 Württembergisches) No.19	9. 7.1683	Stüttgart Wiblingen & Ulm	XIII
Ulanen Regiment (König Wilhelm I) (2 Württembergisches) No.20	24. 7.1809	Ludwigsburg	XIX
Ulanen Regiment (Kaiser Wilhelm II König von Preussen) (3 Saxon) No.21	1.10.1905	Chemnitz	XIX
1st. Bayerisches Ulanen Regiment (König Wilhelm II von Preussen)	21.12.1863	Bamberg	II Bavarian
2nd. Bayerisches Ulanen Regiment (König)	21.12.1863	Ansbach	II Bavarian

REGIMENTAL FACING, BUTTON AND LANCE CAP COLOURS

Regiment	Facings	Buttons	Tschapska
No.1 Guard	White (Piped red)	White	White
No.2 Guard	Red	Yellow	Red
No.3 Guard	Yellow	White	Yellow
Line Regiments			
1st. Ulans	Red	Yellow	White
2nd. Ulans	Red	Yellow	Red
3rd. Ulans	Red	Yellow	Yellow
4th. Ulans	Red	Yellow	Light Blue
5th. Ulans	Red	White	White

Line Regiments	Facings	Buttons	Tschapskas
6th. Ulans	Red	White	Red
7th. Ulans	Red	White	Yellow
8th. Ulans	Red	White	Light Blue
9th. Ulans	White	Yellow	White
10th. Ulans	Crimson	Yellow	Crimson
11th. Ulans	Yellow	Yellow	Yellow
12th. Ulans	Light Blue piped white	Yellow	Light Blue
13th. Ulans	White	White	White
14th. Ulans	Crimson	White	Crimson
15th. Ulans	Yellow	White	Yellow
16th. Ulans	Light Blue piped white	White	Light Blue
17th. Ulans	Crimson	Yellow	White
18th. Ulans	Crimson	Yellow	Crimson
19th. Ulans	Red	White	Red
20th. Ulans	Yellow	Yellow	Yellow
21st. Ulans	Crimson	White	Light Blue
Bavarian Regiments			
1st. Bavarians	Crimson	Yellow	Crimson
2nd. Bavarians	Crimson	White	Crimson

The 2nd. Prussian Guard Ulan Regiment on parade. Note that the officers have Guard Stars on the front and rear of their shabraques.

REGIMENTAL COLONELS-IN-CHIEF

Regiment	Colonel-in-Chief
1st. Guard	—
2nd. Guard	King Albert of Saxony prior to 19.6.1902 and succeeded by King Frederick August III on 15.10.1904.
3rd. Guard	—
1st.	Michael Alexander, Grand Duke of Russia.
2nd.	General of Cavalry von Hanisch.
3rd.	Serge, Grand Duke of Russia.
4th.	General of Cavalry, Frederick William George Ernest of Prussia
5th.	Grand Duke Adolf of Luxemburg, Duke of Nassau.
6th.	—
7th.	Grand Duke of Baden.
8th.	1899-1906 Prince Albert of Prussia. 5.10.1908 King Alfonso XIII of Spain.
9th.	Grand Duker Frederick William of Mecklenburg Strelitz.
10th.	Archduke Franz Ferdinand of Austria-Este.
11th.	General of Cavalry, Count von Haeseler.
12th.	—
13th.	The Kaiser.
14th.	Archduke Joseph of Austria.
15th.	—
16th.	Prince George of Saxony.
17th.	Emperor Franz Joseph of Austria.
18th.	Prince Albert of Prussia, Regent of the Duchy of Brunswick.
19th.	1. The King of Wurttemberg. 2. Duchess Vera of Wurttemberg Grand Duchess of Russia.
20th.	The Queen of Wurttemberg.
21st.	The Kaiser.
1st. Bavarian	The Kaiser.
2nd. Bavarian	King Otto of Bavaria.

Grossherzog Frederick William von Mecklenburg-Strelitz, Colonel-in-Chief of Ulan Regiment No.9.

Prince Adolf Friedrich of Mecklenburg-Strelitz, in the uniform of Ulan Regiment No.13 (reproduced by gracious permission of Her Majesty the Queen).

A One-Year Volunteer of Ulan Regiment No.10.

SIMILARITIES BETWEEN THE ULANS AND THE DRAGOONS

The following components of the organisation and uniforms of the Ulans were the same as for the Dragoons:-

The ranks and their distinctions
The changes in the titles of ranks which took place in 1899
The Regimental Adjutants
Fencing distinctions
Farriers' distinctions
Officers' summer tunics
Officers' sashes
Swordknots
Trumpets
Lances
Boots

The following components of the Ulan uniform were the same as for the Dragoons with the noted exceptions:-

Trumpet Major/Bandmaster *(Stabstrompeter/Musikmeister):* From 1912, the uniform was enhanced by large new pattern shoulder straps. These were twisted cords and were in the colour of the epaulette field; purple red for the Saxons and crimson for the Bavarians. On the shoulder straps were special silver lyre badges which were worn in addition to the regimental devices. The regimental devices were officers' pattern. On parade, officers' pattern epaulettes were worn with special cording on the edges of the strap part and on the bridles. On the crescents, under the regimental devices, cyphers, or numbers, the silver lyres were worn. The leather of the pouches, the bandoliers and the sword belts and slings were lacquered. Officers' pattern wallets were carried on the horse.
After five years in the rank of Musikmeister the man could be promoted to Obermusikmeister, in which case a thin central gold thread was added to the shoulder straps and the crescent of the epaulettes. The Musikmeister continued to wear the shoulder wings but the lace was now silver for all regiments with vertical bars instead of the earlier diagonal pattern and the bottoms of the bars were point-ended and edged on either side with thread in the state colour. Around the lower edge of the wing was a band of similarly decorated lace from which long fringing was suspended. On parade the old *fanfarentrompette* was carried with cords in the state colours.

Standard-bearer *(Standartenträger):* The bandoliers of the 1st. and 2nd. Prussian Guard Regiments and the 20th. Regiment were crimson; those of the 8th. Regiment were dark blue and those of the 13th., 15th. and 16th. Regiments were poppy red.

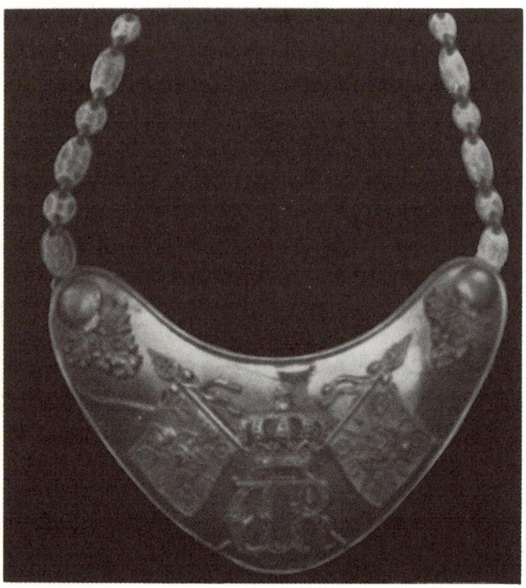

Left: The gorget of the standard bearer of a Prussian ulan regiment.

Right: An ulan standard bearer, taken from a sketch by an Italian artist dated 1909.

One-year Volunteer *(Einjährig-Freiwillige):* The distinguishing badge was a twisted cord around the strap part of the epaulette. This cord was in the state colours and followed the same colours worn by Dragoons.

Re-enlisted Men *(Kapitulanten):* The distinguishing badge was an epaulette bridle in the state colours. For the 19th. Regiment the bridle was in reversed colours, that is to say, black with a central red stripe (instead of the usual Württemberg red with a black stripe) because the regiment had red facings.

Sharpshooters' insignia *(Schützenabzeichen):* The cords were worn on the right side of the *ulanka* from the epaulette bridle to the second button.

Riding School insignia *(Reitschulenabzeichen):* The distinguishing badge for one completed course was a twisted cord, in yellow for regiments with red facings. It was otherwise red, worn across the epaulettes immediately on the crescent side of the epaulette bridle. A soldier who had attended two courses wore a flat piece of worsted material in the same colours and the same position. Until 1905 ulans of the Saxon Regiments who had attended the Riding School at Dresden wore a short rectangular piece of silver lace above the pointed cuff.

Telegraphists' insignia *(Militär-Telegraphenabzeichen):* The distinguishing badge was a piece of worsted braid in the state colours at the button end of the epaulette strap fitting between the metal side strips and on the button end of shoulder straps following the line of the edges.

Prussian and Saxon ulan officers with cuirassiers. They are all wearing the uberrock and feldhinde.

Frock Coat *(Überrock):* The coat was very dark blue for the Prussian Regiments and the Württemberg units, light blue for the Saxons and green for the Bavarians. Regiments with light blue facings had white piped collars, cuffs, and pockets. The Bavarian frock coat had seven buttons in each row down the fronts, red collars piped green and red pipings down the edge of the coats besides the cuffs and pockets.

Working Tunic *(Litewka):* The front and lower edge of the collar, the edge of the front of the coat and the top edge of the cuffs were piped with the colour of the "ulanka" pipings. The collar tabs were in the epaulette field colour and in the case of the 12th. and 16th. Regiments (regiments faced with light blue) the collar tabs were piped all round with white. The Saxon officers had crimson tabs with white piping. The officers of the 21st. Regiment had cornflower blue tabs. The Bavarian officers had plain green litewkas. For N.C.O.'s and troopers the collar patches were in the regimental facing colour.

Officers' Greatcoats *(Mantel/Paletot Umhang):* For the Prussian and Württemberg Regiments, the collars were dark blue when folded down and the facing colour when turned up. In the case of the regiments faced with light blue there was a white piping along the collar edge which showed when the collar was folded down or turned up. The collars for Saxon officers were light blue piped white when folded down and the Bavarians had green collars.

Greatcoat for N.C.O.'s and troopers *(Mantel):* The collar patches were in the regimental facing colour and the shoulder straps in the colour of the epaulette field. The 17th., 18th. and 21st. Regiments had cornflower blue shoulder straps piped white with white numerals. The Prussian Guard Regiments had two bars of white Garde-litzen on each collar patch. The woven cyphers and numerals on the shoulder straps were in yellow on red and crimson material, but red in all other cases.

Fatigue Jacket *(Drillisache):* The Corporal had the lower edge of the collar trimmed with a twisted cord in the state colours. Sergeants had the collar edged with a flat braid in the state colours and One-year Volunteers had the shoulder straps trimmed with twisted state coloured cord. The Sergeant-Major had the sergeants' pattern braid around the lower arm.

Trousers *(Langetuchhosen):* The trousers of the Prusfu and the Württemberg Regiments were blue/black with red welts down the outside seams. The 10th. and 14th. Regiments had crimson pipings. Saxon Regiments wore light blue trousers with crimson stripes along the outside seams, and Bavarian Regiments wore dark green trousers with facing colour stripes.

Riding Breeches *(Reithosen):* The breeches were plain dark blue-black with black leather booting on the insides of the legs for all Prussian and Württemberg Regiments. The Saxons wore light blue breeches with broad red stripes and the Bavarians plain green breeches. For parade dress the Saxon officers had a narrow white welt between the red stripes.

Above: Three views of the ulan überrock. *Below left:* Archduke Ferdinand of Austria in the uniform of a Prussian ulan, wearing the übberock. *Below centre:* Crown Prince Wilhelm wearing the Bavarian überrock. *Below right:* The Württemberg überrock.

Above left: Tschapska of a trooper of a Guard ulan regiment (Tradition). *Above right:* Tschapska of an N.C.O. of a Guard ulan regiment with special pattern cords and tassels and paraderabatte (Tradition). *Left:* Two views of the tschapska of a line ulan regiment. The National cockade was worn on one side only.

LANCE CAP *(Tschapska)*

The 1889 model lance cap had a crown of highly polished lacquered black leather for all ranks. The cap was comparatively low if compared with the British lancer cap of the period. The "Mortar-board" top had a narrow stem connecting it to the crown. The sides of the stem were covered with a detachable, ribbed cloth (the *Paraderabatte*) in the colour of the field of the epaulettes, for parades and ceremonial occasions. The cloth cover was clipped over the four sided swelling stem, but its flat top was left as exposed black leather. The angles and the edges of the cover were trimmed with a button colour braid for officers and with a white piping for N.C.O.'s and troopers. On the left front angle of the cap a large oval cockade in the state colours was clipped to the top edge. For officers the white parts of the cockade were in silver cord. The Prussian regiments had black and white cockades, the Bavarians light blue and white, the Saxons green and white, and the Württembergers red and black.

Chin scales were brass for all regiments irrespective of the button colour; the exception being the 21st. (Saxon) Regiment and the 2nd. (Bavarian) Regiment. Under the boss of the scales on the right side of the cap was a large round metal cockade in the National German colours, that is to say, black, white and red. The officers also wore this ornament but their

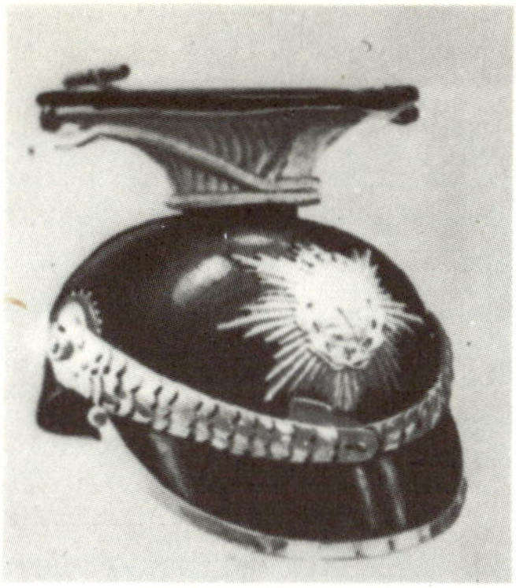

Above left: Tschapska of a trooper from a Bavarian ulan regiment. *Above centre:* Saxon ulan officers' tschapska. *Above right:* Prussian officers' tschapska showing parade plume and cords. *Near right:* Officer of a Guard ulan regiment. *Far right:* Leutnant of Ulan Regiment No.9.

cockades were enamelled with white areas silvered and deepetched.

A drooping hair plume was worn from a socket at the rear of the cockade for parades and ceremonial occasions. The plume was white for troopers and mixed with black for officers and N.C.O.'s of the Prussian regiments. Officers and N.C.O.'s of Bavarian Regiments had light blue and white plumes and the Saxons and Württembergers had green and white and white, black and red respectively.

All trumpeters had red plumes as ustinction of their appointment. General Officers had plumes of aigrette feathers. The small leather visor on the front of the cap was trimmed with button colour metal.

All Prussian regiments had the heraldic eagle as cap plates, Bavarians had the State Arms, Wurttembergers also had the Arms of their State and the Saxons the State Arms on a Star. All helmet plates were in the metal of the colour of the regimental buttons and lace. Officers of the 3rd. Regiment had the added distinction of a band of gold lace around the crown of the lance cap just below the root of the stem.

Regulations specified that the Tschapska should be worn at all times with the front visor 3cm. over the right eyebrow and 5cm. above the left eyebrow. In other words, worn cocked to the right. The chin scales were worn well under the chin.

Helmet Plates: The 1st., 2nd. and 3rd. Guard Regiments had the 'Guard Eagle' with the Star of the Order of the Black Eagle on its breast. The 1st. and 2nd. Line Regiments had the 'Line' eagle with an oval plaque bearing the cypher 'FWR' on its breast. Beneath this was a scroll with the inscription *Mit Gott Für Köenig Und Vaterland* thereon. The 3rd. and 4th. Line Regiments had the 'Line' eagle as for the 1st. and 2nd. regiments, but with the letters 'FR' on its breast beneath the motto. The 13th. Regiment had the 'Guard Eagle' with Star as for the three Guard Regiments, plus a motto scroll over the Star with the honours 'Peninsula'/'Waterloo'/'Garcia-Hernandez', thereon. The 14th. Regiment helmet plate was as for the 3rd. and 4th. Regiments, but with the honour scroll of the 13th. Regiment. The 17th., 18th. and 21st. Regiments had the Saxon armorial Star and arms on their helmet plates. The helmet plates of the 19th. and 20th. Regiments bore the Arms of Württemberg. The 1st. and 2nd. regiments had the Arms of Bavaria.

Cap lines were used with the lance cap. These were retained by a toggle or hook at the top of the right angle. The cap lines (*fangschnur*) then passed around the stem of the cap and fell to the body and were looped up around the collar of the jacket where they passed around the body, with the tassels and flounders hanging on the breast.

Cap lines were white for other ranks and N.C.O.'s and silver cord mixed with state coloured silks for the officers. N.C.O.'s had a worsted line in the state colours worked into the cord, tassels and flounders. The Sergeant-Majors of the Prussian Guard Regiments had special cords in black, silver and white mixture with silver and black tassels and flounders. The white 1884 pattern manoeuvre cover was replaced by a 'reed green' cover in 1892.

Ulan helmet plates: (Photographs courtesy Imperial War Museum.)

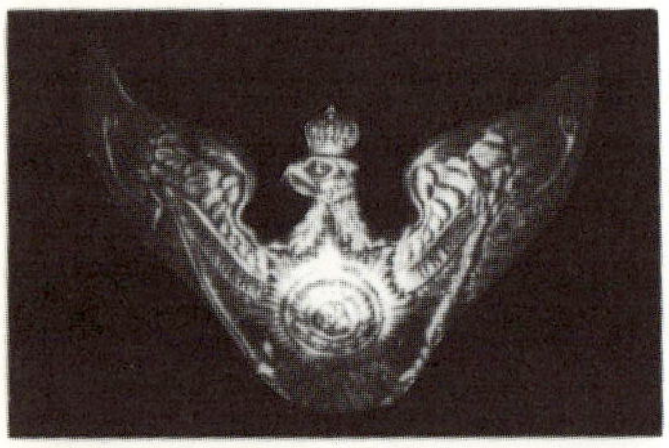

Other ranks, Guard ulans.

Line regiments Nos. 1 and 2.

Regiments Nos. 19 and 20.

Officers, Guard Ulans.

Regiments Nos. 17, 18 and 21.

1st. and 2nd. Bavarian.

Near right: An officer of a Prussian Guard regiment wearing the manöverkappe over his tschapska. *Far right:* A wachtmeister of a Prussian Guard regiment wearing the special pattern tschapska cords and tassels.

UNDRESS CAP *(Mütze)*

The Ulans' undress cap was blue for Prussians and Württembergers, white for Saxons and green for Bavarian Regiments. The cap band and the piping around the crown were in the facing colour. The 12th. Regiment had a white piping along the top edge of the band. The three Saxon Regiments wearing the white cap had cap bands in light blue, red and light blue respectively. On the cap band was a small round metal cockade in the State Colours and directly above this on the crown of the cap a similar cockade in the National colours of black, white and red. Officers and senior N.C.O.'s had the cap fitted with a small black visor.

FULL DRESS TUNIC *(Ulanka)*

The Ulan full dress tunic was a double breasted garment with a row of seven buttons either side of a plastron front. The rear of the tunic had two buttons at waist level and two vertical pockets, one each side of the central vent. Each pocket flap had three buttons. There was a further button at the point of each "Polish" cuff. The tunic was blue for Prussian and Württemberg Regiments, pale blue for the Saxons and a mid-green for the two Bavarian units.

In accordance with an order dated 2.2.1854, the coat was fitted with a detachable plastron front worn only for parades and on ceremonial occasions. This plastron (*paraderabatten*) was in the regimental facing colour except for the 1st. Prussian Guard Regiment, in which case it was white, piped with red. Collars (cut square and hooked close

Above left and centre: Front and rear views of the ulanka of a trooper of the 2nd. Guard Ulan Regiment. (Imperial War Museum). *Above right:* Ulanka of a Sergeant of a line ulan regiment (Imperial War Museum).

for the Prussian Guard Regiments and Regiments Nos. 17, 18, 19 and 21, and round fronted for the remainder), and cuffs were also in the regimental facing colours.

The tunic was piped with facing colour (except Regiments 12, 16, 17 and 18 in which cases the ulanka was piped white) along the sleeve seams, the pockets, the bottom edge of the coat, and down the front in line with the outer edges of the plastron, except in the Bavarian Regiments in which case the piping down the right skirt front was omitted. When the plastron was not worn the pipings traced the shape on the breast of the coat.

On each shoulder was a small reinforced hole through which the hook on the underside of the epaulettes was thrust after being passed under a broad bridle, or loop, of cloth in the facing colour which was sewn on the shoulders of the coat. In some cases the epaulettes were secured by screw buttons in patent holes in the coat.

The collars of the Ulanka were piped as follows:

Regiment	Collar piping	Regiment	Collar piping	Regiment	Collar piping
1st. Guard	none	7th.	blue	16th.*	white
2nd. Guard	none	8th.	blue	17th.*	light blue
3rd. Guard	none	9th.	blue	18th.*	light blue
1st.	blue	10th.	blue	19th.	blue
2nd.	blue	11th.	blue	20th.	blue
3rd.	blue	12th.*	white	21st.*	light blue
4th.	blue	13th.	blue	1st. Bavarian	green
5th.	blue	14th.	blue	2nd. Bavarian	green
6th.	blue	15th.	blue		

* White piping on the plastron also.

Above left: Ulanka of an officer of Ulan Regiment No.9 (Imperial War Museum). *Above centre:* Ulanka of an unteroffizier of the 3rd. Guard Ulan Regiment (Imperial War Museum). The Wachtmeister's ulanka above has the Dresden riding school badge.

On each side of the collar the three Prussian Guard Regiments and Regiment Nos. 13, 17, 18, 19 and 21 wore two bars of Garde-litzen in the button colour, silver or gold for officers, and white or yellow wool for N.C.O.'s and troopers. N.C.O.'s wore only one bar of this lace on their collars because of the restricted width left by the edging rank lace. A bar of the Garde-litzen was also worn on each pointed cuff. Each bar had a light of facing colour running through its centre.

The N.C.O.'s wore the distinguishing badges of their rank, that is to say, broad strips of metallic lace along the top and front of the collar and along the edges of the cuffs. Sergeants and Lance-Corporals were distinguished by buttons with heraldic devices on either side of the collar in line with the epaulette button. The Sergeant's button was larger than the Lance-Corporal's. The Permanent Staff Regimental Sergeant-Major wore an additional band of metallic lace around the lower sleeve 7mm. above each cuff.

EPAULETTES *(Epauletten)*

A detachable epaulette was worn on each shoulder of the ulanka. They had a cloth strap in the colour of the paraderabatten of the tschapska except for the Saxon Regiments who wore all metal epaulettes. The cloth strap was edged on each side with a thin strip of metal scaling. The epaulette shoulder strap terminated at the sleeve end in an oval area (the field) edged with a large metal crescent. The crescent was brass for other ranks and gilt for officers, except for the 1st. Prussian Guard, 3rd. Prussian Guard and the 13th., 15th. and 20th. Regiments all of which had white metal, (silver for officers). The 21st. Regiment had epaulettes wholly of white metal. The regimental number or device was worn within the oval field of the crescent in yellow metal. (see table). Field Officers and General Officers had bullion fringing to the epaulettes. The cloth bridle through which the tongue or frog of the epaulette passed was also in the same colour cloth as the epaulette field. For officers, the bridle was silver thread worked with stripes in state coloured silks. The epaulettes on the left shoulder were furnished with a small brass hook through which the pouch belt was fastened to avoid it slipping.

Regiment	Epaulette Field And Bridle	Regiment	Epaulette Field And Bridle	Regiment	Epaulette Field And Bridle
1st. Guard*	White	7th.†	Yellow	16th. †	Light Blue
2nd. Guard	Red	8th.	Light Blue	17th.‡	Light Blue
3rd. Guard*	Yellow	9th.	White	18th.‡	Light Blue
1st.	White	10th.	Crimson	19th.†	Red
2nd.	Red	11th.	Yellow	20th.*	Yellow
3rd.	Yellow	12th.	Light Blue	21st.‡	Light Blue
4th.	Light Blue	13th.*	White	1st. Bavarian	Crimson
5th.†	White	14th.†	Crimson	2nd. Bavarian	Crimson
6th.	Red	15th.*	Yellow		

For service dress with the tunic, frock coat, working tunic and summer tunic and with the two types of greatcoat officers wore corded shoulder straps (*Achselstücke*) in lieu of the epaulette (for details see the Dragoon section under Rank Distinctions). The shoulder cords were stitched on a lining of cloth the colour of the epaulette field.

Far left: An ulan major showing the fringed epaulettes. *Near left:* Colonel Hans von Mangoldt of Saxon Ulan Regiment No.17. *Below:* Troopers' epaulette of Ulan Regiment No.14 (Imperial War Museum).

* Silver crescents for officers, N.C.O.'s, Trumpeters and Troopers.
† Silver crescents for officers, but brass for all other ranks.
‡ All metal epaulettes, bridles and lining only in Light Blue.

EPAULETTE AND SHOULDER STRAP DISTINCTIONS

Regiment	Distinctions
1st. Guard	Plain
2nd. Guard	Plain
3rd. Guard	Plain
1st.	Cypher 'A III' surmounted by a crown; in gilt for officers, yellow metal for other ranks.
2nd.	Numeral '2'; gilt for officers, yellow metal for other ranks.
3rd.	Cypher 'A II' surmounted by a crown; gilt for officers, yellow metal for other ranks.
4th.	Numeral '4'; gilt for officers, yellow metal for other ranks.
5th.	Numeral '5'; gilt for officers, yellow metal for other ranks.
6th.	Cypher 'CR IX' surmounted by a crown; gilt for officers, yellow metal for other ranks.
7th.	Numeral '7'; gilt for officers, yellow metal for other ranks.
8th.	Numeral '8'; gilt for officers, yellow metal for other ranks.
9th.	Numeral '9'; gilt for officers, yellow metal for other ranks.
10th.	Numeral '10'; gilt for officers, yellow metal for other ranks.
11th.	Numeral '11'; gilt for officers, yellow metal for other ranks.
12th.	Numeral '12'; gilt for officers, yellow metal for other ranks.

Ulan epaulettes:

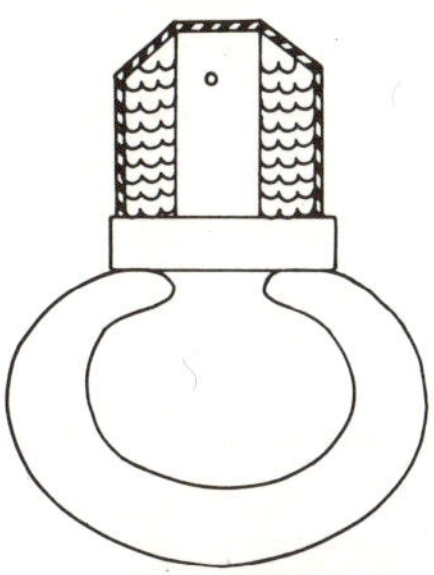

Einjährig Freiwillige

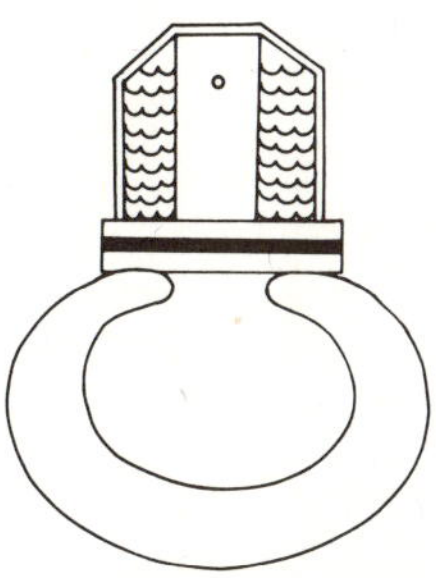

Kapitulant

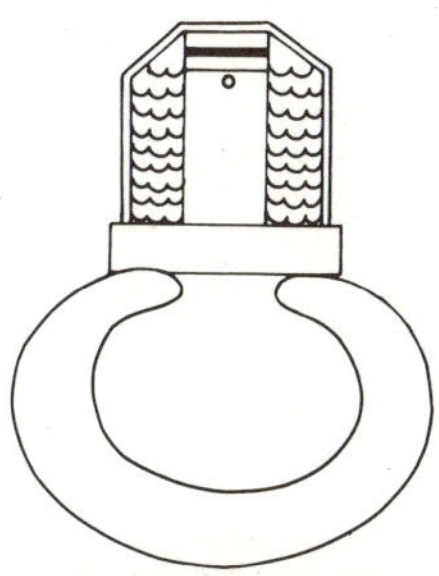

Militär Telegraph

Two year courses Reitanstalt (white braid strip), Regiment No.18

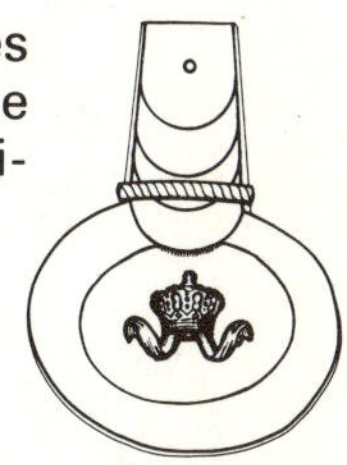

Reitschulenabzeichen 1 year

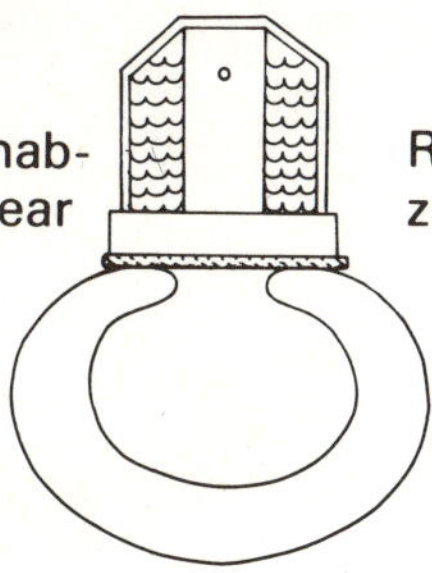

Reitschulenabzeichen 2 years

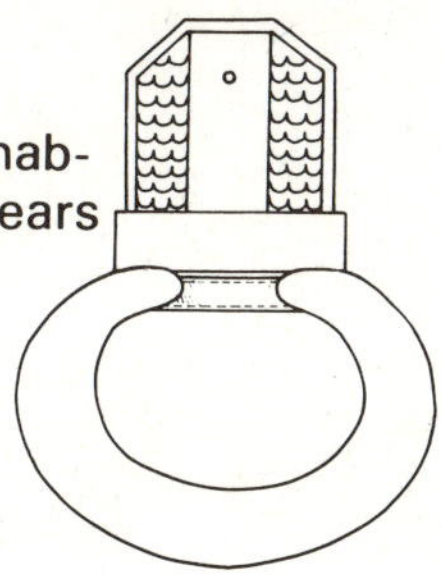

Kapitulant (green stripe in white braid), Regiment No.18.

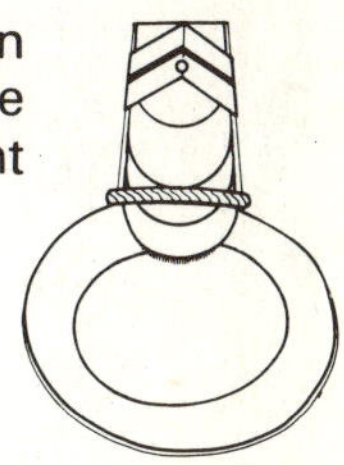

Regiment	Distinctions
13th.	Cypher 'WR II' surmounted by a crown; gilt for officers, yellow metal for other ranks.
14th.	Numeral '14'; gilt for officers, yellow metal for other ranks.
15th.	Numeral '15'; gilt for officers, yellow metal for other ranks.
16th.	Cypher 'GR' surmounted by a crown; gilt for officers, yellow metal for other ranks.
17th.	All yellow metal epaulettes with the Imperial German Crown within the crescent for all ranks.
18th.	Plain yellow metal epaulettes for all ranks.
19th.	Cypher 'KR' surmounted by a crown; gilt for officers, yellow metal for other ranks.
20th.	Cypher 'W II' surmounted by a crown; gilt for officers, yellow metal for other ranks.
21st.	White metal epaulettes for all ranks with the Imperial German Crown within the crescent.
1st. Bavarian	Plain
2nd. Bavarian	Plain

Other ranks' shoulder strap and epaulette distinctions:

Regiment No.1

No.6

No.13

No.16

No.19

No.20

GALA TROUSERS *(Galahosen)*

In Gala dress officers wore dark blue trousers with a piping down the seams and a broad cloth stripe on each side of the piping. For the 1st. Prussian Guard Regiment the wide stripes were white and the centre piping red. In the 12th. and 16th. Regiments the broad stripes were light blue and the piping white. For all other regiments, both the piping and stripes were of the facing colour.

GIRDLE *(Passegurtel)*

The girdle was worn around the waist over the ulanka. It was 2⅜ inches wide and made of leather covered with cloth coloured equally in thirds. The centre strip was blue and the two outer thirds the facing colour. The 12th. and 16th. Regiments had the added distinction of a narrow piping of white on the outer edges of the two outer light blue stripes. The Bavarian Regiments had white waistbelts with plain brass open buckles in lieu of the girdle.

OFFICERS' UNDRESS WAIST-BELT *(Feldbinde)*

These belts were introduced in 1896, and were worn for orders of dress when the sash was not worn. The belts were approximately 5cm. wide and were made in silver woven thread with silk stripes worked in. They were closed by a circular locket and clasp at the front and had slides either side of the clasp similar in colour to the belt. The colours for the various contingents are detailed below.

Prussia: A silver belt with two black stripes, the locket and clasp silver or gold according to the button colour of the regiment. The locket design was the cypher "WRII" with a crown over, the clasp encircled in a closed wreath of laurel leaves.

Bavaria: A silver belt with two pale blue stripes, the locket and clasp silver or gold according to button colour. The locket design was the Bavarian crown, the clasp encircled in a closed wreath of laurel leaves.

Saxony: A silver belt with two pale green stripes, the locket and clasp in silver or gold according to button colour. The locket design was a Gothic 'A' with a crown over, the clasp encircled in a closed wreath of laurel leaves.

Württemberg: A silver belt with two black/red stripes, the locket and clasp silver or gold

Below left: An ulan of Saxon Ulan Regiment No.17 showing the girdle. *Below right:* Graf Zeppelin wearing the überrock with the undress waist-belt, in conversation with the Kaiser.

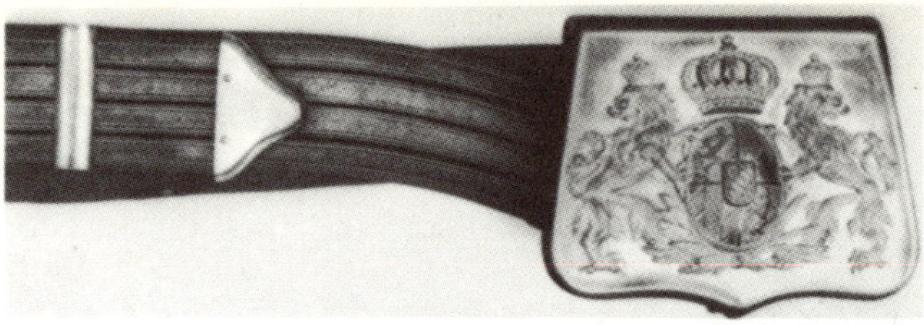

Left: An officer of the 2nd. Guard Ulan Regiment reporting to the Kaiser. Note the pouch with the Guard Star emblem. *Above:* A Bavarian ulan officer's pouch-belt and pouch (Wallis and Wallis). *Below:* Silver pouch for Saxon ulan officers.

according to button colour. The locket design a Gothic 'W' with a crown over, the clasp encircled by a closed wreath of laurel leaves.

The Bandmaster wore a special Moroccan leather belt faced with the regimental colour and edged with N.C.O.'s lace of the button colour. The locket and clasp were silver or gold according to button colour and the lockets were decorated with a lyre and the clasps encircled by a closed wreath of laurel leaves.

POUCH AND POUCH-BELT *(Kartusche/Bandelier)*

The Ulans had white leather pouch belts worn over the left shoulder retained under the epaulette strap by a small brass hook on the epaulette itself. Officers of Prussian Regiments had blue Moroccan leather belts faced with lace in the button colour. The officers of the 1st. Prussian Guard Regiment had red Moroccan leather pouch belts trimmed with silver lace. The pouch lids of all ranks of the three Guard Regiments had the Star of the Order of the Black Eagle on the lid. The 13th., 19th. and 20th. Regiments all had the 'WR' cypher on the pouch lid. All the remaining Prussian Regiments had the 'FWR' cypher on their pouches. Pouches and belts of the Bavarian officers were red Moroccan leather, the face of the belt trimmed with silver lace with three light blue silk lines along its face. The lids of the pouches were covered with silver plate and bore the Bavarian Arms. On the other hand the pouches of the Bavarian rank and file had plain black lids. The inevitable exception to the rule were the trumpeters who had a special plate with the Arms of Bavaria superimposed with a Star.

The Saxon Officers also had pouches and belts of red leather trimmed with silver lace. On

Officers' sabres:

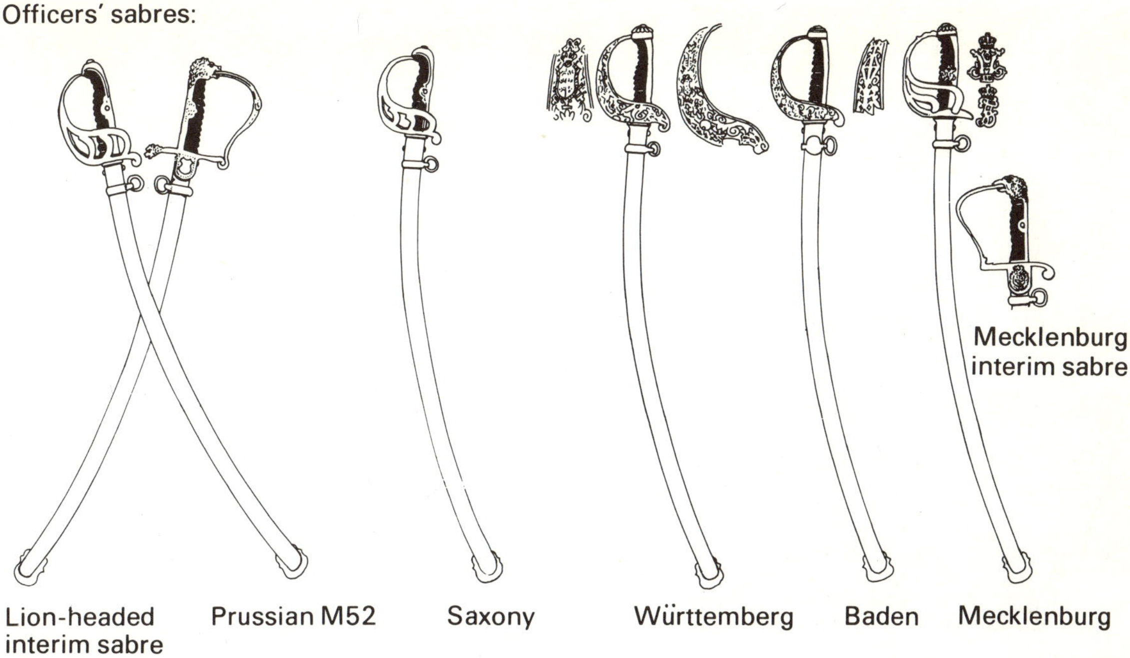

the fronts of their belts were gilt shields and crowns surrounded with wreaths. Their pouch lids were gilt plated and decorated with the Arms of Saxony, the edges of the lid embellished with Saxon trefoils. Saxon troopers had plain black pouches and the trumpeters a brass plate consisting of the Saxon shield surmounted by a crown and enclosed in a wreath.

SWORDS *(Säbel)*

Until 1873, all Prussian Ulans carried the M52 sabre. From that date they were issued with French 'Chasseur' sabres. These were mounted with stirrup bow guards and black leather grips bound with steel wire. The scabbards were steel and had two bends to support the rings. In 1889 the Ulans began to take into use the M89 weapon which had a pierced steel guard and a deeper pommel, and a straight blade. Prussian officers, N.C.O.'s and Trumpeters did not take the French sabres into use but continued to carry the M52 until the 1889 issue became available to all. Officers of the 1st. Regiment and some N.C.O.'s continued to carry old French pattern sabres of the 1812-1815 period.

Saxon lancers carried the heavy cavalry sword with steel guard and Saxon arms. Saxon officers carried swords with gilded pommels and the lion's head sabres for undress duties.

Württembergers carried swords with the State Arms on the pommel and the officers swords with guards in the so-called Austrian pattern except for undress when a light sword with a stirrup guard and with the lion's head pommel was used.

Bavarians had swords with the Bavarian Arms on the guards.

PARADE SHABRAQUE *(Paradeüberlegedecke)*

The shabraques were worn only for parades and ceremonial occasions. They were dark blue for all Prussian and Wurttemberg Regiments and green for the Bavarian Regiments. The shabraque was cut with rounded corners both at the front and the rear and was trimmed with a broad band of worsted braid in the facing colour. Along the edge was a further piping of the same colour. The regiments with light blue facings had a white piping along either side of the worsted band. The Prussian Guard Regiments and the 13th. Regiment had the Guard Star in the rear corners in button colours. The officers had the Star in the front corners also and in their case the Stars were all worked in silver embroidery with the centres finished in coloured silks.

The Saxon Regiments had black sheepskin shabraques. Their officers had silver stars in the rear corners of the shabraques and in addition hexagonal crimson housings trimmed with silver lace for the 17th. and 21st. Regiments, and gold lace for the 18th. The lace had three red lines worked through its length. The Bavarian shabraques had crimson stripes and piping and crimson Bavarian Crowns in the rear corners. Ulan officers' also used the dark blue undress "interims" shabraque.

An ulan officer (foreground) on parade with shabraque.

WEAPONS, BRIDLE, SADDLERY AND PERSONAL EQUIPMENT *(Waffen/Reitzeug)*

Prior to 1908 the M79 revolver was carried by cavalry but was replaced by the M08 automatic (Luger) pistol in that year. The M98, or M88, carbine was carried in a bucket on the right side of the saddle, except for the Bavarian cavalry in which case it was carried in a case fixed to the off wallet and resting on the rider's thigh. Saxons carried a Sharps .433 calibre weapon.

The saddle was similar to the British army saddle of the period, but lighter and higher in the arch, while the flaps were deeper and wider. The girth buckle protector flaps, and the panels were numnah. A thick woollen blanket, 7 feet 7 inches by 6 feet 4 inches, was folded four times and placed under the saddle. The stirrups were wide and heavy and made of nickelled metal. The girths were of the Cape pattern and a surcingle was carried. The wallets were much more roomy and larger than the wallets of a British cavalry saddle, and more in the nature of small saddle bags. The nearside wallet contained a pocket for cartridges, and the shoe case was on the offside. A spade or hatchet could also be attached. Bits also differed to the British army pattern, the bridoon being thicker, and the big bit had a considerably lower port and shorter cheek.

A short piece of rope with a ring at one end was carried over the carbine bucket. The rope was fastened, together with others carried by the rest of the troop, or squadron, to form the picketing line in bivouac. The corn bag was shaped like a sausage and was carried strapped to the rear of the saddle over the mantle.

A mess tin made of aluminium, holding 4½ pints was carried in a leather case on the near-side of the saddle. The tin cover could be used as a cup or small frying pan, and a separate handle for the cover fitted inside the mess tin. The waterbottle was carried on the person as with the infantry. When fully equipped in marching order, the weight carried by a troop horse was about 100 lb., minus man.

An officers' saddlery and equipment consisted of regulation saddle, blanket (*Woilach*), two wallets, one saddle bag, shoe case, greatcoat cover, corn sack, drinking bucket and leather head rope.

At the outbreak of war in 1914 cavalry regiments acting with cavalry divisions were provided with telephone and bridging equipment and also with explosives and demolition equipment.

STANDARDS *(Standarten)*

All Ulan regiments carried standards, with the exception of the Saxon regiments (17th., 18th. & 21st. regiments). They were all either rectangular or cut with two points in the design known as the 'Dragonenmuster', except the two Württemberg regiments (19th. and 20th.) who had a special 'Vexillum' pattern. The staff was approximately 3 metres tall and painted. The pike head was gilded and from its base a number of ribbons and cravats were suspended. These streamers are detailed below.

1900 Streamers were awarded by Kaiser Wilhelm II to all his regiments in 1900. The streamer was in the State colours and had two gilt clasps attached to it. One clasp had on one side the Imperial Crown and on the reverse the date '1 Januar 1900'. The other showed the monograms of the various royal houses surmounted by a crown, and on the back the two dates: "1 Januar 1900" with the date of the raising of the regiment below this. In June 1913, to celebrate his Silver Jubilee the Kaiser presented all his 'Life' Regiments with a second pair of clasps, in this case silver, which were attached to the 1900 streamer above the previously mentioned gilt clasps. These had on one side the crowned monogram or cypher 'WRII' and on the reverse '1888/15 Juni/1913'.

1813/15 Campaign streamer was orange, with thin white and black stripes on either edge.

1848/49 Campaign streamer was black with a white stripe at each edge. Awarded with or without golden swords.

1864 Campaign streamer was black with an orange stripe on one edge and white stripe on the other. Awarded with or without golden swords.

1866 Campaign streamer was black with white and orange stripe along each edge. Crossed swords in gilt at the bottom for most regiments taking part.

1870/71 Campaign streamer had the black and white ribbon of the Iron Cross with the addition of a central red stripe. Decorated with golden clasps inscribed with the names of the various actions in which the regiments took part. This streamer was awarded in 1895 upon the Silver Jubilee of the campaign.

ULAN STANDARDS AND STREAMERS

Types of Standard are illustrated and lengthy descriptions are unnecessary but some points are described for clarification.

1st. Guard Ulans: Square type. White field with black corners. Centre orange with a light blue scroll. Silver embroidery with gold monograms and fringe.

Streamers: 1866, KDM 1870/71, 1900. Standardbearer's bandolier white with silver lace.

2nd. Guard Ulans: As for the 1st. Guard Regiment including streamers. Bandolier red with gold lace.

3rd. Guard Ulans: Square type. Field of white shot silk. Centre and corner motifs red shot silk. Embroidery and fringe silver.

Streamers: As 1st. and 2nd. Guard Regiments. Bandolier yellow with silver lace.

1st. Ulans: Square type. All black with the corners traced with gold. Orange centre with light blue scroll. Embroidery and fringe gold.

Streamers: 1813/15, 1866, KDM 1870/71, 1900, Centennial streamer. Standardbearer's bandolier red with gold lace.

2nd. Ulans: Swallow-tailed type. Crimson field with white corners edged black. Centre white with a red scroll. Embroidery and fringe gold.

Streamers: As for the 1st. Regiment, but no swords on the 1866 streamer. Standardbearer's bandolier red with gold lace.

3rd. Ulans: Square type. Black field with white corners. Orange centre with light blue scroll. Embroidery and fringe gold.

Streamers: As for the 1st. and 2nd. Regiments and in addition parade streamers presented by Empress Maria of Russia in 1857 and the streamer of the Russian Order of St. George presented in 1876. Bandolier red with gold lace.

4th. Ulans: Swallow-tailed type. Light blue field, white corners edged with black. White centre, light blue scroll. Gold embroidery and fringe.

Streamers: as for 1st. Regiment but no Centennial streamer. Standardbearer's bandolier red with gold lace.

5th. Ulans: Swallow-tailed type. White field. White corners edged with black. White centre with blue scroll. Silver embroidery and fringe.

Streamers: As for the 4th. Regiment. Bandolier red with gold lace.

6th. Ulans: Swallow-tailed type. Red field with white corners edged black. White centre, red scroll. Silver fringe and embroidery.

Streamers: 1813/15, 1849 with swords, 1866, KDM 1870/71, 1900, Centennial streamer. Bandolier with silver lace.

7th. Ulans: Swallow-tailed type. Yellow field with white corners edged with black. White centre, red scroll. Silver fringe and embroidery.

Streamers: As for the 6th. Regiment, but no Centennial streamer. Bandolier red with silver lace.

8th. Ulans: Swallow-tailed type. Light blue field with white corners edged black. White centre, light blue scroll. Silver fringe and embroidery.

Streamers: As for the 7th. Regiment. Bandolier light blue with silver lace.

9th. Ulans: Square type. Black field with white corners. Orange centre with black scroll. Gold fringe and embroidery.

Streamers: 1866, KDM 1870/71, 1900. Bandolier white with gold lace.

10th. Ulans: As for the 9th. Regiment including streamers. Bandolier crimson with gold lace.

11th. Ulans: As for the 9th. Regiment but with the addition of the 1864 streamer with swords. Bandolier yellow with gold lace.

12th. Ulans: As for the 9th. and 10th. Regiments. Bandolier light blue with gold lace.

13th. Ulans: As for the 9th. Regiment.

Streamers: KDM 1870/71, 1900 streamer with the additional clasps awarded for Kaiser Wilhelm's Silver Jubilee in 1913 and with a parade streamer presented by Kaiser Wilhelm in 1899 to mark the 10th anniversary of his Colonelcy-in-Chief. Bandolier white with silver lace.

14th. Ulans: As for the 9th. Regiment.

Streamers: 1870/71, 1900 streamer. Bandolier crimson with silver lace.

15th. Ulans: As for the 14th. Regiment, including streamers. Bandolier yellow with silver lace.

16th. Ulans: As for the 14th. Regiment. Bandolier light blue with silver lace.

17th., 18th., and 21st. Ulans (Saxon): Saxon regiments did not carry standards.

19th. Ulans (Württemberg): Special pattern. Square 'Vexillum' type as the Garde du Corps pattern. Dark red field with the Arms of Wurttemberg on the front, i.e. a black rampant lion

and gold (yellow) stag as supporters standing on a light blue scroll. On the reverse side, a gold cypher. Gold fringes and embroidery. (See 25th. Dragoon Standard).
Streamers: KDM 1870/71, 1900, Centennial streamer. Bandolier red with silver lace.
20th. Ulans (Württemberg): As for the 19th. Regiment. Streamers as above but with the addition of the Military Distinguished Service Order (Militär Verdienst Orden). Bandolier yellow with gold lace.
1st. Bavarian Ulans: Square type with a white field. Light blue corners, in the centre the Bavarian Arms surrounded by a green wreath. Gold cypher in each corner surrounded by green wreaths. Gold outer trim with silver fringe.
Streamers: 1900 streamer and a parade streamer presented by Kaiser Wilhelm II in 1892. In addition the colours bore the Bavarian Cross (on its ribbon) for the 1866 campaign and also the Prussian medal for the 1870 campaign. Bandolier crimson with silver lace.
2nd. Bavarian Ulans: As above but no parade streamer. Bandolier crimson with gold lace.

DRUM BANNERS

5th. (Westphalian) Regiment: Awarded 1875.
7th. (Rhenish) Regiment: Awarded 1827. Silver drums which were a gift of the Grand Duke Karl of Saxe-Weimar-Eisenach on the 26.5.1827.

The banners comprised six upper full panels bearing the heraldic Prussian eagle on a shield surmounted by a crown and surrounded by laurel wreaths. The lower panels comprised five full panels and two half panels bearing the crowned monogram 'FWR'

Parade drum banners of Ulan Regiment No. 7.

surrounded by laurel wreaths. The border, fringes, tassels, monograms and crowns were embroidered in gold, the shields silver and the laurels in silver and green. The ground colour of both tiers was white.
13th. (Hanovarian) Regiment: Awarded 1899. This regiment inherited the silver drums of the old Hanoverian Garde du Corps. The drum banners had an upper and lower tier of panels. The top panels were of silver material bearing the Imperial Eagle and 'WR' monogram. The lower panels were of blue. All decorated with silver lace. Fringes and tassels of silver. The 'interim' drum banners were blue with silver lace, and the panels were decorated with a star and monogram alternately. Like the trumpeters, kettledrummers wore an all red plume.
1st. Bavarian: Awarded 1909. The upper panels were dark green. Upon these were placed oval medallions surrounded by gold foliage and containing the following emblems: the Prussian heraldic eagle on a white shield with sprays of gold oak and laurel leaves on each side and surmounted by the Prussian crown; the monogram WRII surmounted by the imperial crown, all in gold, and surrounded by the chain and badge of an order; the Arms of the King of Bavaria on a shield surmounted by a gold crown and surrounded by sprays of gold embroidered oak and laurel leaves. All the panels terminated in gold fringes and tassels. The upper edge of the banner was trimmed with gold lace from which gold tassels were suspended on long gold cords.

The lower panel was entirely crimson (the facing colour) and its straight lower edge was trimmed with a gold fringe.

Regiment	Trumpeters	Kettledrum horse where known
1st. Guard	Dark Browns	
2nd. Guard	Chestnuts	
3rd. Guard	Mixed: Chestnuts, Browns and Blacks	
1st.	Dark Browns	
2nd.	Chestnuts	
3rd.	Chestnuts	
4th.	Browns	
5th.	Chestnuts	Chestnut
6th.	Chestnuts	
7th.	Blacks	Black
8th.	Blacks	
9th.	Light Browns	
10th.	Browns	
11th.	Blacks	
12th.	Chestnuts	
13th.	Chestnuts	Chestnut
14th.	Chestnuts	
15th.	Dark Browns	
16th.	Chestnuts	

3: Bavarian Chevaulegers

Besides the two regiments of ulans the Bavarian Army had eight regiments of Chevauleger (light horse). These light cavalry regiments were originally raised in 1790. In 1811 there were six regiments and in 1905 the 7th. was raised. The 8th. was raised in 1909.

Basically dressed as ulans, the fundamental difference in the uniform was the type of head-dress which was worn. For the Chevauleger this was the leather pickelhaube and not the tschapska. The uniform tunic and trousers were a peculiar colour known as *stahlgrun* (steel green). In effect this was a rather dark green tone with an olive-green cast.

Army Corps	Title	Raised	Garrison
III Bavarian	1st. Chevauleger Regiment Kaiser Nikolaus v. Russland	1682	Nürnberg
III Bavarian	2nd. Chevauleger Regiment Taxis.	1682	Regensberg
II Bavarian	3rd. Chevauleger Regiment Herzog Karl Theodore	1724	Dieuze
I Bavarian	4th. Chevauleger Regiment König	1744	Augsburg
II Bavarian	5th. Chevauleger Regiment Erzhzg Friedrich v. Österreich	1776	Sarrgemund
III Bavarian	6th. Chevauleger Regiment Prinz Albrecht v. Preussen	1803	Bayreuth
III Bavarian	7th. Chevauleger Regiment Prinz Alfons	1905	Straubing
I Bavarian	8th. Chevauleger Regiment	1909	Dillingen

Bavarian chevaulegers wearing manöverkappen over their helmets.

REGIMENTAL COLONELS-IN-CHIEF (1902)

1st. Regiment	Czar Nicholas II of Russia
2nd. Regiment	Albert, Prince of Thurn and Taxis.
3rd. Regiment	Duke Karl Theodor of Bavaria
4th. Regiment	Otto, King of Bavaria
5th. Regiment	Frederick, Grand Duke of Austria
6th. Regiment	Prince Albert of Prussia, Regent of the Duchy of Brunswick
7th. Regiment	Prince Alphonso of Spain
8th. Regiment	Vacant

REGIMENTAL FACING AND BUTTON COLOURS

Regiment	Facing Colour	Button Colour
No. 1	Crimson	Yellow
No. 2	Crimson	White
No. 3	Pink (Peach)	Yellow
No. 4	Scarlet	White
No. 5	Scarlet	Yellow
No. 6	Pink (Peach)	White
No. 7	White	Yellow
No. 8	White	White

Officers' Uniforms: For parade dress the officers wore epaulettes with crescents in the button colour, the straps and field of the epaulettes was in the facing colour and the strap part edged with State coloured lace. The bridles to the epaulettes were silver with two light blue stripes and the metal crescents were in the button colour. Field officers had fringing suspended from the crescents. For all orders of dress officers wore red Moroccan leather pouch belts faced with silver and with three light blue lines and the red leather showing at the edges. Their pouches were also red leather with silver plated lids with the Bavarian Arms. The sash was silver with light blue stripes and the undress waist belt followed the same pattern. In undress the epaulette was replaced by shoulder cords darted with light blue.

The officers' frock coat was green with a collar in the facing colour piped with green and with facing colour piping on the cuffs and the pockets. The piping was also worn down the front edge of the coat and it had two rows of seven buttons. The officers' overcoats and the cloak were trimmed with dark green collars which were in the facing colour when turned up. The officers' housings were plain green, rectangular, and were worn under the saddle. The officers' working tunic was plain green and cut similar to the Bavarian ulans.

Headdress was a black leather spiked helmet (*pickelhaube*) for all ranks with a square front visor and a detachable metal spike which could be exchanged for a white falling hair plume for parades and ceremonials in the same manner as for Dragoons. The cap plate comprised

A chevauleger officer wearing the umhang.

A group of chevaulegers photographed in 1914. Note the brown leather belts and the tunics piped on one side only.

the Bavarian State Arms and the cap had metal chin scales with the Bavarian State cockade under the left hand boss and the German 'national' under the right. All metal fittings and the cap plate were in the regimental button colour.

The undress cap (*mütze*) was green with a cap band and piping in the facing colour and with the State and National cockades. The officers and the N.C.O.'s had small black visors on this cap.

Tunic: Although similar to the Ulanka it differed in a number of significant details. The shoulder straps were pointed and were in facing colour cloth and the coat had round cuffs and collars in the facing colour. There was no piping in the back seams or the sleeves although the rear vent and the pockets were piped in the facing colour. The bottom edge of the coat was also piped in this fashion. A detachable plastron in the facing colour was worn on the coat for parades and ceremonial occasions and when this was removed the coat edge with its plastron shaped top was traced with facing piping. Unlike the ulanka this piping did not trace out the left hand edge of the plastron area. Each round cuff had two buttons. The buttons retaining the shoulder straps were decorated with the regimental number and the edges of the collars were piped with green for all ranks.

Breeches: Plain green for all ranks.

Walking out Trousers: Green with a wide band of the facing colour down the outside of each leg.

Belts: Pouch belts and waist belts plain white with open brass buckles on the waistbelt and a plain black pouch. The lance girdle *(passegurtel)* was not worn.

N.C.O.'s and Trumpeters: Button colour braid on the collars and cuffs. Heraldic buttons in the regimental colour where appropriate. Trumpeters had wings in the facing colour decorated with lace in the button colour.

Overcoats: Grey with facing colour collar patches and shoulder straps. N.C.O.'s had distinguishing grade strips of white with pale blue stripes.

Shabraques: Dragoon pattern with a wide worsted braid band and piping in the facing colour and with Bavarian Crowns in the rear corner in the facing colour.

Lance Pennons: Light Blue and white; N.C.O.'s plain white with the Bavarian Arms.

Trumpet Cords: Light Blue and white.

The Bavarian chevauleger tunic.

THE BAVARIAN CAVALRY STANDARDS

The Bavarian chevauleger and ulan regiments did not possess standards until 1882. In that year, however, the Chevauleger Regiments 1 to 4 were issued with laid-up standards of the former Garde du Corps and 2nd. Cuirassier Regiments. The basic design of all these standards was similar, having on one side the crowned and mantled royal arms surrounded by a wreath and on the other the crowned monogram MJK, also encircled by a laurel wreath. That awarded to the 1st. Chevauleger was light blue with gold embroidery and fringes (ex Garde du Corps), that of the 2nd. was light blue with silver embroidery and fringes (ex Garde du Corps), that of the 3rd. was also light blue and gold (ex 2nd. Cuirassiers), and that of the 4th, light blue and silver (ex 2nd. Cuirassiers).

At the same time, the 5th. and 6th. Chevaulegers and also the 1st. and 2nd. Ulans received new standards manufactured according to the pattern of the 1841 issued infantry colours, but of smaller dimensions and surrounded by a narrow embroidered gold border and, on the three open sides, with silver fringes. In 1906 the 7th. Chevaulegers (raised in 1905) were issued with a standard of this type with border and fringes of gold, and in 1910 the 8th. Chevaulegers (raised in 1909) received a similar one but with border and fringes of silver. However, these two later issued standards had a corner motif consisting of a crown in place of the crowned and interlaced L of the earlier models. The hafts of all the standards were light blue.

The Bavarians decorated their standards both with actual campaign medals and, in some cases, with centennial or parade streamers. In addition, the regiments in existence in 1900 received the commemorative streamer awarded by Wilhelm II to all his regiments to mark the turn of the century. In the case of the Bavarians it was half light blue and half white and the gilt clasps bore on one the monogram LPR in place of the WRII of Wilhelm II and on the other the date 18 Januar 1871 in place of Januar 1900. The table below shows the decorations carried on the standards of each of the chevauleger and ulan regiments:-

1st. Ulan Regiment	**C**	**D**	**E**	**G**		
2nd. Ulan Regiment	**D**	**D**	**E**			
1st. Chevauleger Regiment	**A**	**C**	**D**	**E**	**F**	
2nd. Chevauleger Regiment	**A**	**C**	**D**	**E**	**F**	
3rd. Chevauleger Regiment	**A**	**C**	**D**	**E**	**F**	
4th. Chevauleger Regiment	**A**	**C**	**D**	**E**	**F**	
5th. Chevauleger Regiment	**A**	**B**	**C**	**D**	**E**	**F**
6th. Chevauleger Regiment	**A**	**C**	**D**	**E**	**F**	
7th. Chevauleger Regiment	—	—	—	—	—	—
8th. Chevauleger Regiment	—	—	—	—	—	—

Legend A: Cross for the 1815 campaign (a bronze cross on a white ribbon edged with narrow stripes of black and light blue). **B:** Cross for the 1849 campaign (a bronze cross on a red ribbon with narrow green edges). **C:** Cross for the 1866 campaign (a bronze cross on a white ribbon with narrow stripes of light blue towards each edge). **D:** The 1870 Franco-Prussian War medal (a bronze medal on a black ribbon edged with white and with a red central stripe). **E:** The 1900 streamer described above. **F:** Centennial streamer. **G:** Parade streamer.

4: Landwehr and Reserve Officers

The Prussian and Saxon, etc., contingent landwehr officers wore blue tunics (*waffenrock*) with red collar and pointed cuffs and red piping down the leading edge of the coat and on the pocket flaps in the rear of the skirts. The buttons were gilded. The edge of the red collar was piped blue and the collars and cuffs were decorated with a broad flat gold lace edging was piped on either side with a tiny red piping and had on the inner side a gold lace cord with tracing eyes. Under the buttons on the pointed cuff the tracing cord was finished in a small tight Hungarian knot. The landwehr officers of the Guard had a single gold Gardelitzen on each side of the collar and another on each cuff instead of the knot. The officers wore epaulettes or shoulder cords with fields in the colour designated for their Army Corps and wore gilded Arabic numerals of that Corps on both ornaments.

The helmets were the Dragoon pattern with gilded mountings, square cut front visor, cross form plate at the base of the spike and a silver Maltese Cross known as the *Landwehrkreuz* over the centre of the eagle's breast. For the Guard, the Gardeadler was worn with the cross in gold on a silver Guard Star. On parade a white, black for the Guard, hair plume was worn. The blue uberrock had red pipings and the paletot, etc., had the same collars as the infantry, that is to say, red when turned up and blue when folded down.

The riding breeches, boots and belts were all as for Dragoon officers. For Gala wear the Landwehr officers wore dark blue trousers with a gold lace stripe down each outer leg.

Their pouches were worn on blue Moroccan leather bandoliers faced with gold lace. The pouch lids were decorated with crowned Cyphers and the Guard had the Guard Star with the Cross in the centre on their pouches.

Bavarian Landwehr officers wore the uniform of their parent unit with the Army Corps

Landwehr and Reserve officers helmet plates showing the position of the Landwehr cross. *Below left:* Prussian line regiments. *Below right:* Guard regiments.

Mecklenburg

Oldenburg

Saxony

Hesse

Baden

Bavaria

Württemberg

number on the epaulettes and shoulder cords and the Landwehr Cross on the helmet plates.

Officers on the reserve wore regimental uniform distinguished by the cross on the breast of the eagle helmet plate and the rank and file of reserve formations were similarly distinguished. When the manöverkappe was worn it was painted or appliqued with a large red 'R' in the front.

THE DISTINGUISHING COLOURS OF LANDWEHR OFFICERS' SHOULDER ORNAMENTS

Army Corps.	Epaulette field and shoulder cord underlay	Army Corps.	Epaulette field and shoulder cord underlay
I	White	XIII	Red
II	White	XIV	Red
III	Red	XV	Red
IV	Red	XVI	Yellow
V	Yellow	XVII	Yellow
VI	Yellow	XVIII	Light Blue
VII	Light Blue	XIX	Dark Blue
VIII	Light Blue		
IX	White	**Bavarian Army Corps**	
X	White	I	Red
XI	Red	II	Red
XII	Dark Blue	III	Red

The German sources researched for this list are not entirely clear. Both state: "*Die epaulettefelder sowie achselstückfutter waren bei Garde rot, bie Line von der achselfarbe der betressenden Korps mit goldenen arabischen nummer der Korps. . .*". The difficulty being that the Corps lists referred to is straight forward until one gets to the 14th. Corps where the position is not clear. Nevertheless the position has been logically deduced as being shown.

The Strength of Cavalry Regiments at Mobilisation in 1914*

Regiment of four squadrons (half regiment of three squadrons).
36 Officers including medical, veterinary and Administrative Officers.
63 N.C.O.'s including administrative personnel (unterbeamte).
552 troopers.
73 Medical and transport personnel.
Weapons:- 529 carbines, 90 revolvers, 577 lances.

2 pontoon waggons.
1 telegraph/telephone waggon.
1 medical waggon.
5 pack waggons.
5 provision waggons.
5 fodder waggons.
709 riding horses.
60 drawing horses.

* from German sources.

Reserve Regiment of three squadrons.
27 Officers including medical and veterinary staff.
47 N.C.O.'s (including administrative staff).
416 troopers.
48-51 Medical and transport personnel.
Weapons:- 401 carbines, 63 revolvers and 433 lances.
2 pontoon waggons.
1 telegraph/telephone waggon.
4 pack waggons.
4 provision waggons.
4 fodder waggons.
532 riding horses.
36 drawing horses.

Landwehr Squadron.
6 Officers (including medical and veterinary staff).
15 N.C.O.'s
136 troopers.
2 Medical and transport personnel.
132 carbines, 19 revolvers but no lances.
1 pack waggon.
163 riding horses.
2 drawing horses.

Landsturm Squadron.
6 Officers (including medical and veterinary staff).
15 N.C.O.'s
136 troopers.
1 medical orderly.
19 revolvers and pistols.
132 carbines.
160 riding horses.
No transport.

AN EXTRACT FROM A GENERAL STAFF OFFICER'S NOTE BOOK. BERLIN 1914.

Notes on Strength and Organisation of cavalry troops etc.
Squadron: 6 officers, 163 N.C.O.'s and troopers 178 horses.
Landwehr Squadron when mobilised: 6 officers, 153 N.C.O.'s and troopers.
Regiment with four squadrons: 36 officers, 688 N.C.O.'s and troopers. 769 horses. Staff: 8 officers, 30 N.C.O.'s and troopers. 49 horses.
Half Regiment: comprised 3 squadrons. Strength with staff: 26 officers, 516 N.C.O.'s and troopers. 581 horses.
Reserve Cavalry Regiment: 3 squadrons, 27 officers, 511 N.C.O.'s and troopers. 568 horses. Staff: 7 officers, 18 N.C.O.'s and troopers.
Signals Company of a Cavalry Division: 8 officers, 88 N.C.O.'s and troopers. 98 horses.
Signals Company of a Bavarian Division: 8 officers, 122 N.C.O.'s and troopers. 148 horses.
Remount Depot: 3 officers, 63 N.C.O.'s and troopers. 107 horses, 1 provision waggon and 1 field waggon.
Field Hospital: 9 officers, 51 N.C.O.'s and troopers, 28 horses, 1 equipment waggon, 1 medical supply waggon, 1 pack waggon, to each troop plus 1 ambulance and 1 orderly waggon for record purposes.

RESERVE REGIMENTS AND RESERVE COMPANIES

As the troops were mobilised and organised on a war footing the old Landwehr Reserve and Landsturm Regiments and squadrons were re-mustered and by 1914 the following is an outline of the cavalry formation which resulted:-

Mobilised Reserve Cavalry Regiment of the 1st. Army Corps: The Reserve Cavalry Regiment of the 29th. Landwehr Brigade, the Reserve Squadron of the 8th. Ulan Regiment, the 12th. Ulan Regiment and the Reserve Squadron of the 10th. Jäger zu Pferde Regiment.
Reserve Cavalry Regiment of the 6th. Army Corps: The Reserve Squadron of the 4th. Hussars the 6th Hussars, the 2nd Ulans and the 11th. Jäger zu Pferde Regiment.
The 2nd. Reserve Cavalry Regiment of the 6th. Army Corps: The 1st. Field Cavalry Regiment of the 6th. Corps including 1-3 Squadrons from the Reserve Squadron of the 1st. Cuirassier Regiment, the 8th. Dragoon Regiment, 4th. and 6th. Hussars, the 2nd. Ulans and the 11th. Jäger zu Pferde Regiment.

Bavarian Landwehr and Reserve officers' pickelhaube (Wallis and Wallis).

There were also mobilised Reserved Squadrons of the Army Corps as follows:

The 1st.—6th. Mobilised Squadrons of the Reserve Cavalry of the 1st. Army Corps raised from reserve cadres of the 1st. Dragoon Regiment, 8th. Ulans, 12th. Ulans, 10th. Jäger zu Pferde Regiment, 3rd. Cuirassier Regiment, 1st. Dragoons and 9th. Jäger zu Pferde Regiment.
The 1st. and 2nd. Mobilised Squadrons of the Reserve of the 5th. Army Corps from reserves of the 1st. Jäger zu Pferde Regiment, and the 1st. Ulans.
The 1st. and 2nd. mobilised Squadrons of the 17th. Army Corps from reserve squadrons of 1st. Hussars, and the 4th. Jäger zu Pferde Regiment.
The 1st. to 3rd. mobilised Reserve Squadrons of the 20th. Army Corps from reserve squadrons of the 11th. Dragoons, 4th. Ulans and the 5th. Cuirassiers.

The Reserve Cavalry Company of the Guard Corps was formed from the 2nd. Guard Ulan Regiment. In addition there were **Reserve Cavalry Companies** *(Abteilung)* attached to the following Army Corps:-

2nd. from the 2nd Cuirassier Regiment.
3rd. from the 3rd. Cuirassier Regiment.
4th. from the 7th. Cuirassier Regiment.
5th. from the 1st. Jäger zu Pferde Regiment.
6th. from the 1st. Cuirassier Regiment.
7th. from the 4th. Cuirassier Regiment.
8th. from the 7th. Hussars.
9th. from the 15th. Hussars.
10th. from 13th. Ulans.
12th. (1st. Royal Saxon) from the Guard Cavalry Regiment.
13th. from the 20th. Ulans.
14th. from the 20th. Dragoons.
18th. from the 24th. Dragoons.
19th. from the 18th. Ulans.
1st. Bavarian from 1st. Bavarian Heavy Cavalry Regiment.
2nd. Bavarian from 3rd. Chevaulegers.
3rd. Bavarian from 1st. Chevaulegers.

Other small units raised were **Reserve Cavalry Troops** *(Ersatz-Kav. Züge).*

LANDWEHR SQUADRONS

The mobilised Landwehr were formed in squadrons and attached to the various Army Corps and with cadres raised as follows:-

1st. Landwehr Squadron of the Guard Corps	1st. Guard Ulan Regiment
2nd. Landwehr Squadron of the Guard Corps	1st. Guard Dragoon Regiment
3rd. Landwehr Squadron of the Guard Corps	1st. Guard Dragoon Regiment
1st. Landwehr Squadron of the 1st. Army Corps	Cuirassier Regiment No.3

2nd. Landwehr Squadron of the 1st. Army Corps	Dragoon Regiment No.1
3rd. Landwehr Squadron of the 1st. Army Corps	Dragoon Regiment No.1
1st. Landwehr Squadron of the 2nd. Army Corps	Ulan Regiment No.9
2nd. Landwehr Squadron of the 2nd. Army Corps	Grenadier Regiment zu Pferde No.3
3rd. Landwehr Squadron of the 2nd. Army Corps	Dragoon Regiment No.12
1st. Landwehr Squadron of the 3rd. Army Corps	Dragoon Regiment No.2
2nd. Landwehr Squadron of the 3rd. Army Corps	Ulan Regiment No.3
1st. Landwehr Squadron of the 4th. Army Corps	Ulan Regiment No.16
1st. Landwehr Squadron of the 5th. Army Corps	Ulan Regiment No.10
2nd. Landwehr Squadron of the 5th. Army Corps	Dragoon Regiment No.4
3rd. Landwehr Squadron of the 5th. Army Corps	Ulan Regiment No.10
4th. Landwehr Squadron of the 5th. Army Corps	Jäger zu Pferde Regiment No.1
1st. Landwehr Squadron of the 6th. Army Corps	Hussar Regiment No.4
2nd. Landwehr Squadron of the 6th. Army Corps	Hussar Regiment No.4
3rd. Landwehr Squadron of the 6th. Army Corps	Hussar Regiment No.6
1st. Landwehr Squadron of the 7th. Army Corps	Cuirassier Regiment No.4
2nd. Landwehr Squadron of the 7th. Army Corps	Hussar Regiment No.11
3rd. Landwehr Squadron of the 7th. Army Corps	Ulan Regiment No.5
1st. Landwehr Squadron of the 8th. Army Corps	Hussar Regiment No.7
2nd. Landwehr Squadron of the 8th. Army Corps	Cuirassier Regiment No.8
1st. Landwehr Squadron of the 9th. Army Corps	Dragoon Regiment No.17
2nd. Landwehr Squadron of the 9th. Army Corps	Dragoon Regiment No.17
3rd. Landwehr Squadron of the 9th. Army Corps	Dragoon Regiment No.18
1st. Landwehr Squadron of the 10th. Army Corps	Hussar Regiment No.17
2nd. Landwehr Squadron of the 10th. Army Corps	Hussar Regiment No.17
3rd. Landwehr Squadron of the 10th. Army Corps	Hussar Regiment No.17
1st. Landwehr Squadron of the 11th. Army Corps	Hussar Regiment No.14
2nd. Landwehr Squadron of the 11th. Army Corps	Jäger zu Pferde Regiment No.6
3rd. Landwehr Squadron of the 11th. Army Corps	Jäger zu Pferde Regiment No.2
1st. Landwehr Squadron of the 12th. Army Corps	Saxon Garde Reiter Regiment
1st. Landwehr Squadron of the 13th. Army Corps	Dragoon Regiment No.26
2nd. Landwehr Squadron of the 13th. Army Corps	Dragoon Regiment No.25
3rd. Landwehr Squadron of the 13th. Army Corps	Ulan Regiment No.20
4th. Landwehr Squadron of the 13th. Army Corps	Ulan Regiment No.19
1st. Landwehr Squadron of the 14th. Army Corps	Dragoon Regiment No.20
2nd. Landwehr Squadron of the 14th. Army Corps	Jäger zu Pferde Regiment No.5
1st. Landwehr Squadron of the 17th. Army Corps	Hussar Regiment No.1
2nd. Landwehr Squadron of the 17th. Army Corps	Hussar Regiment No.2
3rd. Landwehr Squadron of the 17th. Army Corps	Jäger zu Pferde Regiment No.4
1st. Landwehr Squadron of the 18th. Army Corps	Ulan Regiment No.6
2nd. Landwehr Squadron of the 18th. Army Corps	Dragoon Regiment No.6
3rd. Landwehr Squadron of the 18th. Army Corps	Dragoon Regiment No.23
4th. Landwehr Squadron of the 18th. Army Corps	Dragoon Regiment No.24
1st. Landwehr Squadron of the 19th. Army Corps	Ulan Regiment No.21
2nd. Landwehr Squadron of the 19th. Army Corps	Ulan Regiment No.18
1st. Landwehr Squadron of the 20th. Army Corps	Cuirassier Regiment No.5
2nd. Landwehr Squadron of the 20th. Army Corps	Cuirassier Regiment No.5

3rd. Landwehr Squadron of the 20th. Army Corps	Cuirassier Regiment No.5
1st. Landwehr Squadron of the 21st. Army Corps	Ulan Regiment No.7
1st. Landwehr Squadron of the 1st. Bavarian Army Corps	1st. Bavarian Schwere Reiter Regiment
2nd. Landwehr Squadron of the 1st. Bavarian Army Corps	4th. Bavarian Chevauleger Regiment
1st. Landwehr Squadron of the 2nd. Bavarian Army Corps	2nd. Bavarian Ulan Regiment
1st. Landwehr Squadron of the 3rd. Bavarian Army Corps	2nd. Bavarian Chevauleger Regiment

5: Adjutants

ADJUTANTS-GENERAL OF THE KAISER AND KINGS

The headdress was the General Officers' pattern pickelhaube with crown of polished black leather, square cut visor and gilded or silvered mounts. The helmet had a tall fluted spike with a cruciform base, gilded chin scales and gilded or silvered helmet plates which differed from State to State. For the Prussians the helmet plate was the Guard Eagle with superimposed silver and enamelled Guard Star. Other States were as follows: Saxony: The Gold Saxon Star with the Saxon shield wreathed and crowned in silver thereon. Bavaria: The full State Coat of Arms with Crowned Lion supporters. Württemberg: The full State Coat of Arms with Crowned Lion and Stag supporters. Baden: No Adjutants-General. Hesse: The Silver Crowned Hessian Lion with Star superimposed, the whole surrounded by a wreath. Mecklenburg: The Golden Sunburst superimposed with the Arms of Mecklenburg. Helmet fittings silver.

For parade dress a swan's feather ornament was screwed on the helmet in lieu of the spike. These feathers were white, underlayed with feathers dyed in the State colours; black for Prussia; light blue for Bavaria; green for Saxony; black and red for the Württembergers; light blue for Mecklenburg and red for Hesse.

The tunic was the waffenrock, blue for all states with scarlet collar and Swedish (round)

The King of Bavaria as Colonel-in-Chief of Ulan Regiment No.19.

The Kaiser as Colonel-in-Chief of Saxon Ulan Regiment No.21.

cuffs. The front edge and the pocket flaps were piped with scarlet. The collar was decorated with two loops of *Kolbenstickerei* in gold wire and there were two similar loops on each cuff. Twelve buttons were set down the front of the coat reaching almost to the bottom of the skirts. Silver bullion epaulettes with silver bullion fringes were worn for parades and on all full dress occasions. For other orders of dress silver twisted silver cords with darts of state coloured silk all on a red ground were worn on each shoulder. The bridles, or *passenten,* sewn on each shoulder of the coat, were silver or gold striped with state colours and were also lined red. The silver field of the crescent part of the epaulette were ornamented with gilded state cyphers and crowns. Special gold aiguilettes were worn on the right shoulder. The woven silver or gold (for Mecklenburg) sash with state coloured stripes was worn over the coat, or alternatively for undress duties the 'feldbinde' was worn. The Mecklenburg lace and embroidery was silver and the collar of the "waffenrock" was decorated with a special oakleaf design as was the three pointed blue cuff flap. Saxon Adjutant-Generals had blue collars and cuffs with special gold lace, and for Hesse, the lace was silver and the tunic had "Brandenburg" cuffs, with three bars of silver lace.

A Personal Adjudant to a member of the royal household.

A Personal Adjudant to a royal person. Major Von Heyden Linden of the 1st. Guard Ulan Regiment.

ADJUTANT-GENERALS "A LA SUITE" OF THE KAISER

The officers wore the same uniform as the Adjutant-Generals, but with silver lace on the collar and cuffs. Helmet fittings were also of silver. The epaulettes were the generals' pattern with gold cyphers, etc., and the silver aiguilettes had gold crowns on the pins.

The Adjutant-General had blue horse furniture comprising five-pointed holster caps, and shabraques which were rounded at the front and pointed at the rear and decorated with one wide and two narrow bands of gold lace. The holster caps and rear corners of the shabraque were further decorated with padded silver embroidered Guard Stars surmounted by silver crowns. The undress horse cloth was blue, rectangular with one wide gold laced border.

FLÜGEL-ADJUTANTEN

Prussia: The generals' pattern helmet was worn with silver mountings and a white hair plume. The tunic was a blue waffenrock with red collar and cuffs and silver lace, epaulettes and buttons. Gold 'WR' and crown cyphers were worn on the epaulettes and shoulder cords. The special woven black-striped silver adjutants' sash was worn over the right shoulder. There were two wide red stripes with a narrow red piping between down the outside of the legs of the trousers and breeches. The 1903 pattern Litewka had red collar patches and piping, but no red breast linings. The uberrock was very dark blue with red collar, pipings and breast linings. The paletot, mantel and umhang had the same collars as the infantry. Epaulette bridles were silver, striped black and lined with red.

Bavaria: Generals' pattern helmet with silver mounts and white hair plumes. The tunic was a pale blue waffenrock with red collar and cuffs with the same type of silver ornaments as the Prussians. Silver epaulettes with fringes and red underlay. The epaulette and silver shoulder cord devices were the gold Bavarian Crown on a silver field. The silver woven sash with two light blue stripes was worn over the right shoulder. The collar and cuffs of the waffenrock were of velvet. The breeches had two red stripes and a narrow piping between down the outside seams of each leg. Epaulette bridles were silver with two pale blue stripes and lined with red.

Saxony: Generals' pattern helmet with gold mounts and a white hair plume. The waffenrock was blue, faced with blue, with a red piping along the top and front of the collar, down the front of the coat, along the bottoms of the skirts, along the cuff flaps and top and rear of each cuff. The collar and cuffs were decorated with gold oak leaf embroidery and the buttons were gold. A gold aiguilette was on the right shoulder. The silver epaulettes were worn with silver fringes and the fields were decorated with a gold Gothic 'AR' Cypher with crown and the silver epaulette bridles were striped green and lined with red. Breeches had the outside of each leg decorated with two wide and one narrow red stripe. Flügel-Adjutants to all the Saxon Princes wore the same uniform, but with plain blue cuffs piped red with the collar only decorated with gold oak leaf embroidery.

Württemberg: Dressed as the Prussians but with the State coloured stripes in the sash and on the bridles of the epaulettes and with a 'WR' crowned cypher in gold on the field of each epaulette.

Baden: Dressed as the Prussians, with State coloured stripes in the silver woven sash and epaulette bridles, and gold crowned 'FR' cyphers worn on each epaulette field.

Mecklenburg: Two types of uniform were worn, the 'old pattern' which comprised a blue waffenrock faced red. The red collar had a wide strip of silver lace along the top and front edges, and within this a silver embroidered spray of oak leaves. The cuffs were red with a blue three pointed cuff flap and with three silver buttons. The flap was also decorated with a silver embroidered spray of oak leaves. Silver aiguilettes were worn on the right shoulder. The epaulette bridles were silver with three red stripes, edged black on both sides. No adjutants' sash was worn over the right shoulder, but the golden yellow sash with cornflower blue and red stripes was worn around the waist in lieu. The later pattern waffenrock was as for the Prussians, but with silver *Kolbenlitzen* as decoration. Generals' helmets were worn with silver mountings and white hair plumes.

Hesse: Generals' helmets with silver mountings and white hair plumes with red hair beneath. The tunic was a blue waffenrock faced with red. The collar as for the Prussians, but the cuffs had a blue rectangular cuff flap with three square-ended silver laced bars and three silver buttons. Silver epaulettes with a crowned 'EL' cypher in gold on the field. Bridles silver with three red stripes edged black on both sides. Silver woven adjutants' sash with three red stripes worn over the right shoulder.

The mantel, paletot, umhang and uberrock were as worn by Prussian officers of the 1st. Foot Guards. The mantel and paletot had no breast linings. The 1903 pattern litewka was as for the Adjutant-Generals.

PERSONAL ADJUTANTS TO PRINCES OF ROYAL HOUSEHOLD

A special uniform was worn by officers seconded for adjutantal duties to Princes of the Royal Houses. Although hardly ever worn, regimental uniforms being more the fashion, it is worth mentioning.

They wore General officers' helmets with gilt mounts and a white hair plume as for Flügel-Adjutants. The tunic was a blue, faced red, waffenrock with pointed cuffs. The collar and cuffs were decorated around the edges with a twisted gold lace inside a broad flat gold braid. Ulan pattern epaulettes with gold crescents and a red field and underlay were worn. The bridles were silver with black stripes and had red underlay. On the field of the epaulettes were two gilt stars (Rittmeister rank). The woven silver adjutants' sash was worn over the right shoulder, and silver aiguilettes were also worn from the right shoulder. The helmet carried the Dragoon pattern eagle.

6: Medical and Veterinary Personnel

MILITÄR-ÄRTZE/SANITÄTS-OFFIZIERE (MILITARY DOCTORS)

Until 1873, German Military Doctors were considered to be *Beamte*, or officials and did not strictly rank as officer military personnel. After 1873 they were raised in status and re-titled *Sanitäts-Offiziere*.

Generally the uniform consisted of a blue infantry pattern tunic (waffenrock) with yellow metal buttons and blue facings with the 'Swedish' pattern cuffs. The collar, cuffs, the tunic front and pocket flaps were piped with poppy red. The collar and the cuffs were also decorated with double bars of flat ended, gold lace. The helmet was also the infantry pattern spiked helmet (pickelhaube) and had a rounded front visor, except for the Mecklenburg contingent, in which case the front visor was cut square. The helmet mounts were gilt. The undress mütze was all blue with poppy red piping. The überrock was dark blue with a blue collar, poppy red breast linings and poppy red piping. The mantel, paletot and umhang collars were blue on both sides and piped poppy red. The *Unterärzte* (or junior medical officers) and the *Einjährig-Ärzte* did not wear the mantel, paletot or umhang, but used instead the rank and file mantel. In the General Officers' range the paletot had special carmine pipings. From 1903, the litewka had dark blue collar patches piped with poppy red and poppy red piping to the tops of the cuffs and the front of the jacket. The Unterärzte were wearing the drillisache in 1896, but later took into use the other ranks' litewka with dark blue collar patches piped with red and shoulder straps as worn on the tunic. The doctors were armed with M89 pattern sabre with a black leather, gold trimmed strap and belt. In 1914 a special pattern feldbinde was introduced. This was gold and had two black stripes with an outer trimming of pale blue. The slides were similarly coloured. The clasp was gilded and had the 'W' and crown cypher in the centre and in addition two "Aesculapus" staffs on the surrounding wreath. Epaulettes had a field of dark blue velvet with gold crescents and a gold Aesculapian staff on each field.

STAFF GENERAL DOCTOR OF THE ARMY
(Generalstabsärzt Der Armee)

Generals' pattern spiked helmet with Guard eagle plate with the Guard Star superimposed (the other States used their own special General Officers' helmet plates in lieu). For parade purposes a special swan's feather plume replaced the spike. These were white feathers over an underlay of feathers dyed in the State colours. Epaulettes with stiff gold bullion fringes were worn on each shoulder. The fields of the epaulettes were dark blue velvet, the crescent gold, the edge of the strap was braided with gold, all on a red underlay. The bridles were also under-laid with red. On the field within the crescent was the gold "Aesculapus" staff with a special rosette beneath. Shoulder cocqere a special black, gold and silver design with silver staffs and rosettes thereon. The uniform was completed by the double, red staff pattern, stripes with the narrow red piping between worn down the outside of the trousers and breeches. The General Officers' pattern sword was carried and the coat had twelve buttons down the front terminating almost at the bottom of the skirt.

After raising military doctors from the *Beamte* category in 1873, the following ranks came into being:-

The Generalstabsärzt was granted the rank of Major General.
The General-Ärzt grade had two classes — the 1st. class ranking as colonel and the 2nd. class as Lt. Colonel.
The Divisionärzt and Oberstabsärzt category were considered 1st. class and they ranked as Majors.
The Stabsärzt (a 2nd. class grade) became a Captain.
The Assistenärzt (1st. class) became a lieutenant.
The Assistenärzt Klass II a 2nd. Lieutenant.

All these ranks had the two gold lace bars on the collars and cuffs as previously described. The lower grades known as the Unterärzte and the Einjährigärzte had no lace on the collars and cuffs, and instead of epaulettes and shoulder cords, wore blue shoulder straps edged with a wide silver lace with the State colours in silk running through.

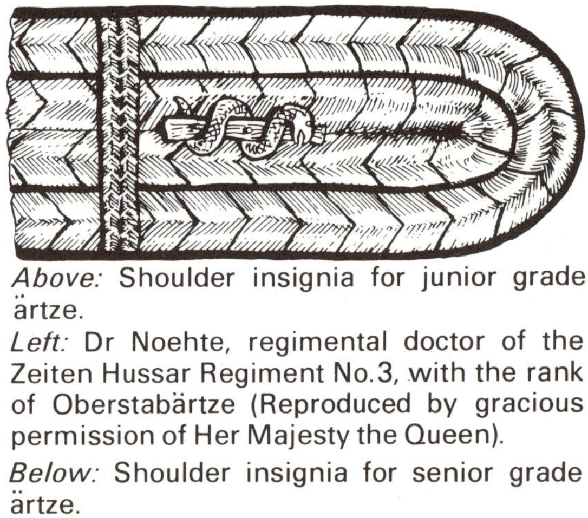

Above: Shoulder insignia for junior grade ärtze.

Left: Dr Noehte, regimental doctor of the Zeiten Hussar Regiment No.3, with the rank of Oberstabärtze (Reproduced by gracious permission of Her Majesty the Queen).

Below: Shoulder insignia for senior grade ärtze.

Tunic for Militär Ärtze

Tunic for Unterärtze

Special Ärtze feldbinde:

Prussia

Baden and Oldenburg

Saxony

Bavaria

Württemburg

Mecklenburg

In February 1914 a special pattern waistbelt (feldbinde) was introduced for the Sanitätsoffiziere. The belt was made of blue Moroccan leather faced with gold lace with stripes in the State colours. The belt and slides were edged with cornflower blue velvet and the gilded locket and clasp with the snake and staff over the laurel leaf wreath except in the case of the Bavarian pattern. For the Saxon Doctors in the General Officer range there were three green stripes on the belt and the slides.

Tunic for Sanitäts Mannschaften from Prussia

Tunic for Sanitäts Mannschaften from Saxony

MEDICAL RANK AND FILE TROOPS *(Sanitäts-Mannschaften)*

From 1867, medical troops had been treated as combatant personnel. They generally wore blue infantry pattern uniforms with blue facings and red pipings and yellow metal buttons. The shoulder straps were also blue (oblong for Saxons but otherwise pointed ended) and edged red and had the Corps numbers in red Arabic numerals. The original grades were as follows:

Lazarett-Gehilfen-Schüler	Private or *Gemeine*
Unter-Lazarett-Gehilfen	Corporal or *Gefreite*
Lazarett-Gehilfen	Sergeant or *Unteroffizier*
Ober-Lazarett-Gehilfen	Staff Sergeant or *Feldwebel.*

Normal N.C.O.'s lace was worn on the collar and cuffs and heraldic buttons were worn on the collars. On the 27.1.1899 the ranks were altered to conform with the new titles of ***Sanitäts-soldaten,*** still graded from private to Staff Sergeant, but the ranks were now preceded by the title 'Sanitäts'. In 1906, the Lazaretten-Sanitäts-Feldwebel of the larger hospitals was granted the appointment of *Spiess* and wore the double band of gold lace around each sleeve and were permitted to wear the uniform of a junior medical officer

(*Unter-Ärzt*), but with blue shoulder straps piped with red. This uniform stood still until 1913 when the Sanitäts personnel attached to cavalry units took into use the uniform of their respective regiments, but without the pouch belt. As a distinguishing badge the 'Aesculapian' staff (which was a yellow snake with red tongue and eyes) was worn on the right upper arm.

However, in the hospitals, the Sanitäts personnel continued to wear the old style uniform, but with red shoulder straps and the Brandenburg pattern cuff with red cuff flaps and yellow metal buttons. On the shoulder straps the Corps number was now in Roman numerals, but in yellow. For the Guard Corps, the medical staff had in addition the Garde-Litzen on the collars and cuffs. Helmets were furnished with brass fittings and the appropriate State helmet plates were worn. Guard units wore, in addition, black horse hair plumes for parades. All the equipment was black leather.

Until 1913 the mantel and litewka had blue collar patches and shoulder straps, the shoulder straps piped red. Boots were the dragoon pattern for the cavalry attachments. In 1888 medical troops attached to cavalry units were given the cavalry pattern swords and belts of the troops to which they were attached (even though still wearing the infantry style of uniform). Sword knots and straps were usually of the pattern issued to the 1st. Squadron. N.C.O.'s carried the normal N.C.O.'s 'Troddel'. The Sanitäts-Feldwebel carried the officers' pattern sword with the officers' pattern sword knot. Sanitäts-personnel attached to the 2nd. Dragoon Regiment were permitted to wear the small distinctive cap eagle on their undress caps, but those attached to the Leib Husaren Regiments did not wear the Death's Head (*Totenkopf*) until the 1913 reforms.

When in the field the white armband with the red cross was worn on the left upper arm.

Stretcher bearers were formed into companies known as Sanitäts-Kompanien. There were three companies to each army corps. They wore a similar uniform to the Dismounted Train, that is to say, blue waffenrock, but with crimson red collars, Swedish pattern cuffs and pipings. White metal buttons. The yellow Corps number was carried on the crimson shoulder straps. The Guard pattern lace was worn where appropriate, with a crimson centre light. Trousers also had crimson pipings. Pistols were carried in holsters and the equipment was as for the Dismounted Train. The mantel had crimson collar patches and shoulder straps. All Saxon units had the oblong shoulder straps.

VETERINARY OFFICERS CORPS *(Veterinär Offiziere Korps)*

An Army Order of the 24th. June 1872 ordered a re-organisation of the veterinary officers corps. For each Army Corps were ***Korps Rossärzte*** (Corps Horse Doctors) and besides these there were ***Stabs-Rossärzte*** (Staff Horse Doctors). From this date the latter grade were known as the Oberrossärzte.

The waffenrock was taken into use by Veterinary staff in 1852. It had black facings and crimson pipings. The Unterrossärzte had black shoulder straps with a gold lace, striped with blue, around the outer edges, while the Rossärzte had a further strip of gold lace down the centre of each strap. The Korps-Rossärzte and Oberrossärzte wore officers' pattern

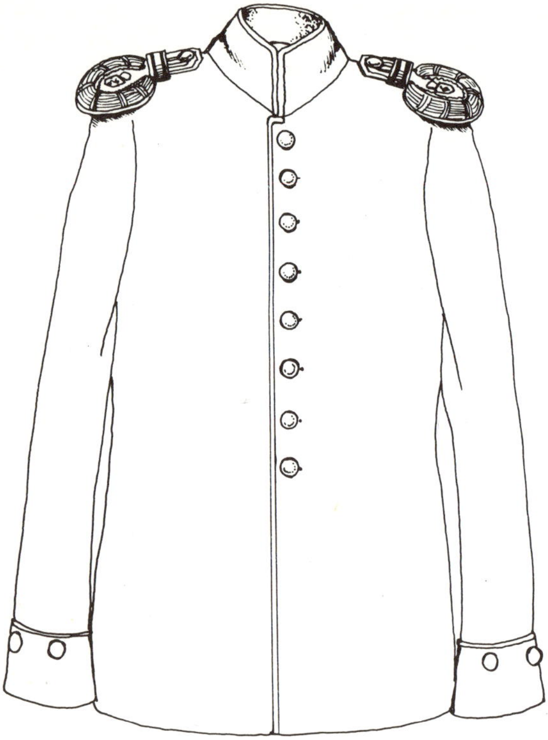

Tunic for Militar Veterinär.

Veterinär insignia.

epaulettes and shoulder cords. The former had epaulettes with black velvet fields and gilt crowned shields bearing the State arms. The underlay of the epaulettes were blue. The lace on the epaulette strap and the bridles were gold striped with blue. The Korps-Rossärzte had, in addition, a golden rosette on the epaulette and shoulder cords from 1888.

The Rossärzte-Eleven wore the "waffenrock" of the 'Unterrossärzte' but without the lace trim on the shoulder straps.

In 1882, the Rossärzte were raised to officer status and were thereafter entitled to wear the epaulettes and shoulder cords. The Oberrossärzte now had one rosette, and the Korps-Rossärzte two. The 'Eleven' was also lifted in status in 1896 and at the same time a new pattern "einjährig" cord was introduced.

In 1903 the Veterinary Corps was further re-organised and the titles of the veterinary officials were thereafter as follows: (1) *Korps-Stabsveterinär*. (2) *Stabsveterinär*. (3) *Ober-veterinär* (4) *Unterveterinär* (5) *Eleven studierende der Militär Veterinär Akademie*.

In 1906 new shoulder ornaments were taken into use. The new epaulettes had crimson fields, linings and bridles. The crescents remained gilt and did not alter. The epaulettes of the Korps-Stabsveterinär were further embellished with an inner and outer silver band. The

Korps and Stabs grades had two rosettes and the Oberveterinär one. The lace trimming on the epaulette strap and on the bridles was silver and blue. For the Korps/Stabs rank the shoulder cords were of a special twisted pattern. The shoulder straps of the Unterveterinär were crimson with a silver and blue lace trimming.

In 1910 the Veterinary officials were once again reformed, this time into a proper Officers' Corps. The titles and grades were once again altered and became as follows: (1) *General-Veterinär* (Colonel). (2) *Korps-Stabs-Veterinär* (Lt. Colonel or Major). (3) *Oberstabs-veterinär* (Major). (4) *Stabsveterinär* (Captain) (5) *Ober-veterinär* (Oberleutnant). (6) *Veterinär* (Leutnant).

Collars and cuffs down to the Stabsveterinär level (Captain) had gold Kapellenlitzen on the collar, cuffs and epaulettes, and shoulder cords as for other commissioned ranks. The fields of the epaulettes, linings and underlay, and bridles still remained crimson, the crescents still gilt. The lace edging to the strap part of the epaulette was still silver, but the woven silk stripe was now in the State colours. The universal rank stars were now taken into use instead of the gilt rosettes. In lieu of the crowned shield badges a gilt snake ornament (Aesculapian) was worn. The epaulettes of the Field Officers had silver and coloured fringing — two strands of silver mixed with two silk strands of the State colour. The shoulder straps of the Unterveterinär had officers' pattern lace with the snake ornaments. Arms for the veterinary officers was the lion's head (cavalry interims) sabre. The Unterveterinär had the Dragoon pattern sabre with an N.C.O.'s troddeln. Until the introduction of the new uniforms in 1910, senior officers wore silver and blue sword knot tassels on a plain black lacquered leather strap. Subsequently the normal officers' sword knot and strap was taken into use. Buttons were yellow metal for Prussia, Saxony, and white metal for Bavaria.

Before becoming part of the Officer Corps the headdress was the spiked helmet, (pickelhaube) of the Beamte pattern. The subsequent helmet had a square cut front visor and a small silver eagle plate on the front. The mütze was blue with a black cap band and crimson piping. The überrock had a black velvet collar piped with crimson, crimson breast linings and all remaining piping crimson. The Unterveterinär did not wear the überrock until 1903, but thereafter the coat was worn with shoulder straps as on the tunic. The mantel, paletot and umhang collars were blue when folded down and black when turned up, piped with crimson. The Unterveterinär wore the shoulder straps as worn on the tunic until 1913. The litewka had black collar patches and crimson pipings. The Unterveterinär wore the rank and file pattern until 1913, without collar patches or shoulder straps. The 'Eleven' wore the rank and file mantel with black collar patches and shoulder straps, and the litewka without patches or shoulder straps. Breeches and boots were as for dragoons, with crimson stripes down the outside seams. *Galahose* of veterinary officers were black with an outer piping of crimson on each side.

7: The Introduction of Field Dress

Two Army Command Orders in 1910 regulated the issue of field-grey (grey-green for some units) field service dress (***Feldbekleidung***) for the German Army and gradually the old full dress uniforms were worn less and less.

The two orders dated 23rd. February and 18th. March of that year prescribed that in future the full dress clothing was only to be worn on special occasions:-

Parade *(Paradeanzug)*
Garrison guard duties *(Garnisonwachtdienst)*
Courts martial duty *(Gerichtsdienst)*
Church parade *(Kirchparaden)*
Walking out dress *(Ausgehanzug)*
On social occasions, balls and festive duties *(Gesellschaftanzug)*

The new field grey uniform was worn for all duties in the field and in training and naturally following mobilisation. The Emperor's manoeuvres in 1910 were probably the last important occasion on which all the units of all the confederated states appeared together in full dress although even then the reed green covers were worn on most of the helmets.

For the Dragoons, the field grey waffenrock had pipings in the facing colours; for the 13th. to 16th. white on the collars and cuffs, and for the 19th. white on the cuffs only. The Guard Lace bars on the collars of the Guard Regiments were worn with square-fronted collars. The Guard officers had cornflower blue collar patches edged with red piping and with silver lace bars. The undress caps had bands and piping in the facing colour and the 2nd. Regiment maintained their traditional eagle badges on these caps.

The ulans succeeded in retaining the ulanka cut for their field grey dress but no longer had the epaulettes with metal crescents. Instead they had shoulder straps buttoned at the collar and swelling out into circular ends at the sleeve to simulate an epaulette form. These had red numbers or cyphers in cloth as worn on the blue full dress clothing. All other pipings on the coat were in the same form as worn on the full dress clothing, except that for the 12th. Regiment the piping on the collar, cuffs and on the trumpeters' wings was light blue instead of white. The Guard Regiments naturally had Garde-litzen as before. The old traditional lance-girdle was not worn with this uniform. The three Guard Regiments had silver Guard Lace on the collars of the officers' ulankas, on white patches for the 1st. Guard Regiment, red for the 2nd. and golden yellow for the 3rd. They had similar bars on the cuffs. Undress caps had band and piping in the facing colour.

8: Cockades

Much has been said in the three volumes of this work on the use of the cockade and the "Nationale" by the various contingents of the "Deutsches Armee". The following list and the illustration elaborates this theme.

The Cockade *(Kokarden):* On the spiked helmet (pickelhaube) the state cockade appeared on the left side. The National appeared on the right. The National cockade had a red centre, a white middle ring and a black outer ring. On the undress cap the state cockade was placed in the front in the centre of the lower band and the "Nationale" on the upper part between the cap band and the piping.

Colours of the "Landeskokarden" in 1900

Prussia	Black centre, white middle ring, outside ring black.
Bavaria	White centre, pale blue middle ring, white outside ring.
Saxony	White centre, green middle ring, white outside ring (the cockade was circular without teeth on the outside edge).
Württemberg	Black centre, red middle ring, black outside ring. Officers red, black outside.
Mecklenburg	Red centre, golden yellow middle ring, cornflower blue outer ring.
Baden	Yellow centre, red middle ring, yellow outside ring.
Oldenburg	White centre, red middle ring, white outside ring.
Hesse	White centre, remainder white but with two red rings (the cockade was circular without teeth on the outside edge).
Brunswick	Pale blue centre, yellow middle ring, pale blue outside ring.

BIBLIOGRAPHY

Krickel-Lange.	Das Deutsche Reichsheer in seiner Neusten Bekleidung und Austrustung.
A. Mila.	Uniformirungs-liste des Deutsche Heeres.
H. Knötel Baron Collas E. Janke.	Friedensuniformen das deutsche heer bei Ausbruch des Weltkrieges.
H. Jurgens.	Uniformen des Deutschen Heeres im Juli 1914.
Moritz Ruhl.	Die Uniformender Deutschen Armee.
War Office.	Handbook of the Ger an Army 1914.
Hochsprug.	Das Deutsche Reichsheer.
Weber.	Deutschlands Armee in Feldgrauer Kriegs-und Friedens-Uniform.
Pietsch.9	Formations—und Uniformierüngsgeschichte. Vol.2.
Rittmeister Ewald Feibig.	Unsterbliche Treue.
	Various issues of the Zeitschrift für Heereskunde.
	Regimental Histories.